ytivitaleR hsifratS

Starfish relativity
Negative numbers do not exist

Martin Erik Horn

Martin Erik Horn
Lilienthalpark
12209 Berlin
Germany

mail@martinerikhorn.de

nroH kirE nitraM
kraplahtneiliL
nilreB 12209
ynamreG

ed.nrohkirenitram@liam

ytivitaleR hsifratS

Starfish relativity
Negative numbers do not exist

Martin Erik Horn

Bibliografische Information der Deutschen Nationalbibliothek:

Die Deutsche Nationalbibliothek verzeichnet diese Publikation
in der Deutschen Nationalbibliografie; detaillierte bibliografische
Daten sind über

http://dnb.dnb.de

im Internet abrufbar.

© 2024 Martin Erik Horn

Verlag: BoD • Books on Demand GmbH, In de Tarpen 42,
22848 Norderstedt

Druck: Libri Plureos GmbH, Friedensallee 273, 22763 Hamburg

ISBN: 978-3-7597-8409-4

stnetnoC

.egnarts railimaf eht ekam ot …"
degnellahc era sevil ruo fo secalpnommoc ehT
".syaw wen ni seussi tuoba kniht nac ew dna

[1] *skcaJ .M ymA & llebA .K ardnaS*
noitacude rehcaet ecneics gnissucsid

dniknaM olleH 0

!sgnieb ralugnatcer olleH !snamuh olleH
,erutcurts lacigoloib ruoy otni detaroprocni era selgna thgiR
.yrtemoeg lanretni ruoy otni

.rennam ralugnatcer a ni gnitaluclac era uoy eroferehT
.snrettap detneiro tfel susrev thgir rof deniart era sniarb ruoy eroferehT
.srebmun evitagen eht detnevni evah uoy eroferehT

.elgna thgir ruoy si sihT
.noitcerid evitisop a otni thgir eht ot gnitniop si smra ruoy fo enO
,noitcerid evitagen a otni tfel eht ot gnitniop si mra dnoces ruoy dnA

.sdrawnwod ro sdrawpu smra ruoy ot ralucidneprep detneiro era sgel dna kcen elihw

.yrtemmys thgir tfel laretalib a evah ton od sehsifrats tnegilletni taht htiw derapmoC
.rennam laretalirt a ni derutcurts era sehsifrats tnegilletnI

.snoitcerid eerht otni tniop yllacirtemmys hcihw smra elcatnet eerht ssessop yehT
.selgna 120° tuoba syawla kniht sehsifrats tnegilletni eroferehT
.srebmun evitagen tnevni ton od sniarb riehT
.scitamehtam evitisop dlofeerht a no desab gnikrow era sniarb riehT

.tsixe ton od srebmun evitageN
.snoitcerid evitisop eerht eht fo hcae rof srebmun evitisop ylno era erehT

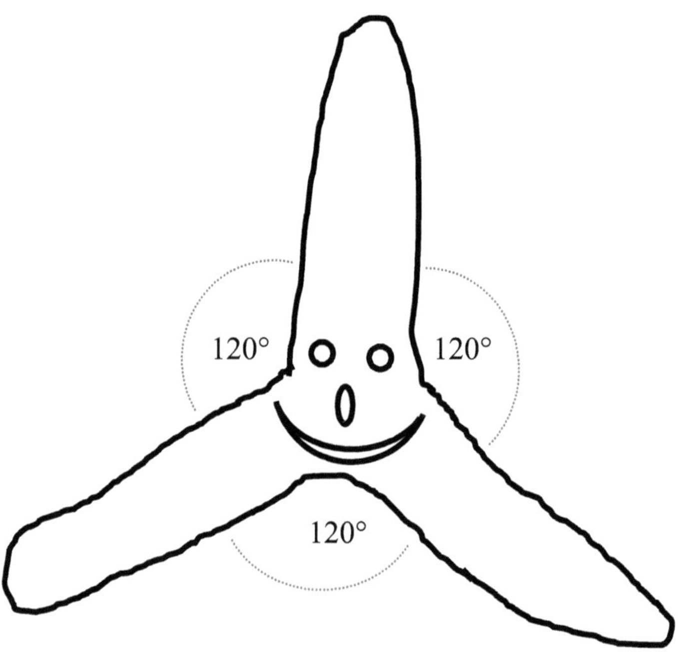

(noititepeR) sehsifratS tnegilletnI fo arbeglA laitapS ehT 1

.tsixe ton od srebmun evitageN
.emit fo htgnel evitagen a derusaem reve sah ,gnieb namuh on ,ydoboN
.ecaps fo ecnatsid evitagen a derusaem reve sah ,gnieb namuh on ,ydobon dnA
relur eht dnuora denrut ylno evah niaga dna niaga su fo llA
.noitcerid rehtona otni secnatsid evitisop derusaem neht dna

.evitisop si ,gnirusaem era ew tahw ,gnihtyrevE
.tsixe ton od srebmun evitageN
.tsixe stnemele evitisop htiw secirtam ylnO

.loohcs hsifrats ta arbegla xirtam nrael sehsifrats elttil eroferehT
noitadnuof eht stutitsnoc secirtam 3 x 3 fo arbegla ehT
.smra elcatnet eerht htiw sehsifrats tnegilletni fo scitamehtam eht fo

.[2] "tsixe ton od srebmun evitageN" koob eht ni denialpxe si arbegla sihT
sehsifrats tnegilletni fo scitamehtam eht fo snoitadnuof cisab eht erehT
.trohs ni ylno ereh taeper lliw ew hcihw ,dnuof eb nac

sehsifrats tnegilletni fo arbegla eht fo snoitcerid eerht ehT
.srotcev tinu eerht yb debircsed era
:yaw gniwollof eht ni deman era yehT

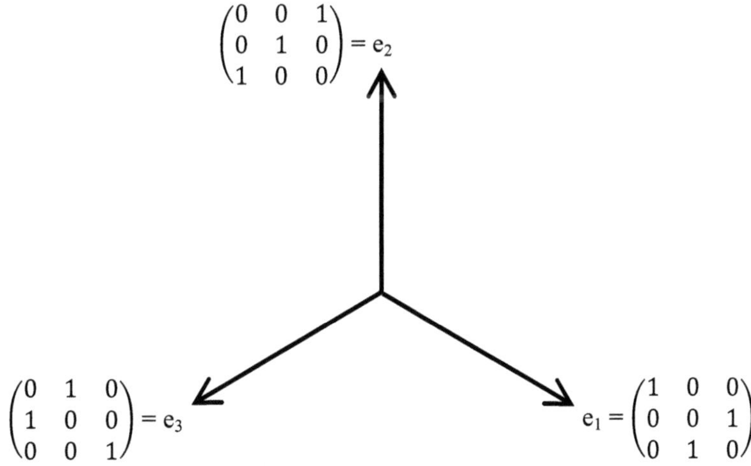

These three unit vectors are pure space vectors.
Thus starfishes have constructed an algebra of two-dimensional space,
as the three unit vectors are lying in a plane.

These unit vectors can be multiplied and added.

A multiplication by itself – squaring – results in the unit matrix or identity matrix.
This identity matrix represents the number one:

$$e_1{}^2 = e_2{}^2 = e_3{}^2 = 1^2 = \begin{pmatrix} 1 & 0 & 0 \\ 0 & 1 & 0 \\ 0 & 0 & 1 \end{pmatrix} = 1$$

All other numbers are multiples of this base unit 1.

All computations are then computations of matrices, e.g.:
Five plus four is equal to nine.

$$5 + 4 = \begin{pmatrix} 5 & 0 & 0 \\ 0 & 5 & 0 \\ 0 & 0 & 5 \end{pmatrix} + \begin{pmatrix} 4 & 0 & 0 \\ 0 & 4 & 0 \\ 0 & 0 & 4 \end{pmatrix} = \begin{pmatrix} 9 & 0 & 0 \\ 0 & 9 & 0 \\ 0 & 0 & 9 \end{pmatrix} = 9$$

And six times seven is equal to two.

$$6 \cdot 7 = \begin{pmatrix} 6 & 0 & 0 \\ 0 & 6 & 0 \\ 0 & 0 & 6 \end{pmatrix} \begin{pmatrix} 7 & 0 & 0 \\ 0 & 7 & 0 \\ 0 & 0 & 7 \end{pmatrix} = \begin{pmatrix} 42 & 0 & 0 \\ 0 & 42 & 0 \\ 0 & 0 & 42 \end{pmatrix} = 42$$

And all other vectors are linear combinations of the three unit vectors.

The most simple linear combination is the null sum:

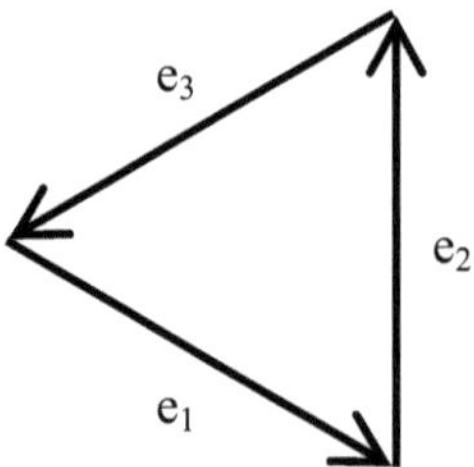

$$e_1 + e_2 + e_3 = \begin{pmatrix} 1 & 0 & 0 \\ 0 & 0 & 1 \\ 0 & 1 & 0 \end{pmatrix} + \begin{pmatrix} 0 & 0 & 1 \\ 0 & 1 & 0 \\ 1 & 0 & 0 \end{pmatrix} + \begin{pmatrix} 0 & 1 & 0 \\ 1 & 0 & 0 \\ 0 & 0 & 1 \end{pmatrix} = \begin{pmatrix} 1 & 1 & 1 \\ 1 & 1 & 1 \\ 1 & 1 & 1 \end{pmatrix} = 0$$

:.g.e ,tsixe orez rebmun eht fo snoitatneserper ynam yletinifni eroferehT

$$2\,e_1 + 2\,e_2 + 2\,e_3 = \begin{pmatrix} 2 & 0 & 0 \\ 0 & 0 & 2 \\ 0 & 2 & 0 \end{pmatrix} + \begin{pmatrix} 0 & 0 & 2 \\ 0 & 2 & 0 \\ 2 & 0 & 0 \end{pmatrix} + \begin{pmatrix} 0 & 2 & 0 \\ 2 & 0 & 0 \\ 0 & 0 & 2 \end{pmatrix} = \begin{pmatrix} 2 & 2 & 2 \\ 2 & 2 & 2 \\ 2 & 2 & 2 \end{pmatrix} = 0$$

ro

$$3\,e_1 + 3\,e_2 + 3\,e_3 = \begin{pmatrix} 3 & 0 & 0 \\ 0 & 0 & 3 \\ 0 & 3 & 0 \end{pmatrix} + \begin{pmatrix} 0 & 0 & 3 \\ 0 & 3 & 0 \\ 3 & 0 & 0 \end{pmatrix} + \begin{pmatrix} 0 & 3 & 0 \\ 3 & 0 & 0 \\ 0 & 0 & 3 \end{pmatrix} = \begin{pmatrix} 3 & 3 & 3 \\ 3 & 3 & 3 \\ 3 & 3 & 3 \end{pmatrix} = 0$$

… cte

$$\Rightarrow \quad \begin{pmatrix} 1 & 1 & 1 \\ 1 & 1 & 1 \\ 1 & 1 & 1 \end{pmatrix} = \begin{pmatrix} 2 & 2 & 2 \\ 2 & 2 & 2 \\ 2 & 2 & 2 \end{pmatrix} = \begin{pmatrix} 3 & 3 & 3 \\ 3 & 3 & 3 \\ 3 & 3 & 3 \end{pmatrix} = \dots = \begin{pmatrix} 0 & 0 & 0 \\ 0 & 0 & 0 \\ 0 & 0 & 0 \end{pmatrix} = 0$$

,orez rebmun eht suht dna xirtam orez a si xirtam a oS
.lacitnedi era xirtam siht fo stnemele lla nehw

.gnihton si oreZ .orez si oreZ
.rotcev orez eht sa gnihton emas eht si orez rebmun ehT

xirtam orez siht ,orez siht srotcev fo noitatneserper dradnats eht ni dnA
.detcelgen eb syawla lliw

,evitisop stnenopmoc eerht eht fo owt ylno syawla era ereht suhT
.noitatneserper dradnats eht ni orez eb lliw tnenopmoc driht eht elihw

si 3 e_1 + e_2 + 2 e_3 rotcev eht fo noitatneserper dradnats eht elpmaxe roF

$$3\,e_1 + e_2 + 2\,e_3 = 2\,e_1 + e_3 + \underbrace{e_1 + e_2 + e_3}_{0} = 2\,e_1 + e_3$$

derongi neeb dah rotcev orez eht retfa
.(egap gniwollof eht no erugif ees)

.srebmun lamron yllaer htiw enod eb nac emas ehT
,srebmun lamron etiuq teg lliw ew dnA

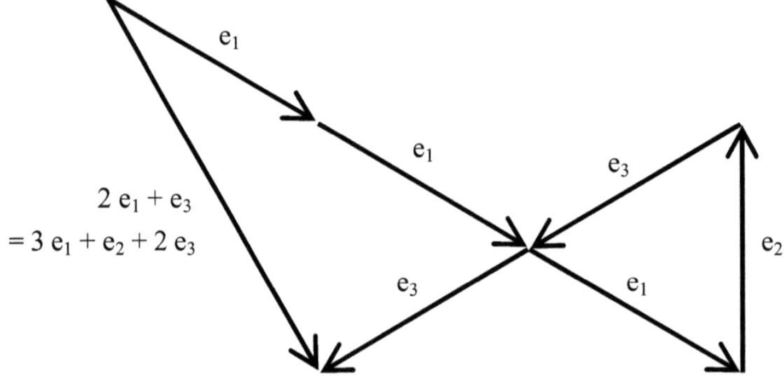

$3\,e_1 + e_2 + 2\,e_3$ rotcev elpmaxe eht ylpitlum ew fi
e_1 rotcev tinu .g.e − rotcev tinu a yb

$$(3\,e_1 + e_2 + 2\,e_3)\,e_1 = 3\,e_1^{\,2} + e_2\,e_1 + 2\,e_3\,e_1 = 3 + e_{21} + 2\,e_{12} = \ldots$$

.srotcev tinu tnereffid owt fo stcudorP cirtemoeG fo tsisnoc smret owt tsal ehT

!srettam srotcev tinu eht fo redro eht ereht dnA
:tnatropmi ylemertxe si srotcaf lairotcev fo redro sihT

$$e_1\,e_2 = e_2\,e_3 = e_3\,e_1 = \begin{pmatrix} 0 & 0 & 1 \\ 1 & 0 & 0 \\ 0 & 1 & 0 \end{pmatrix} = e_{12}$$

$$e_2\,e_1 = e_3\,e_2 = e_1\,e_3 = \begin{pmatrix} 0 & 1 & 0 \\ 0 & 0 & 1 \\ 1 & 0 & 0 \end{pmatrix} = e_{21}$$

.tnereffid era stcudorp owt ehT

ecaps laitaps ylerup ,lanoisnemid-owt eht nI
.gninaem cirtemoeg gnicnivnoc a evah stcudorp eseht

,shtgnel edis lacitnedi evah hcihw smargolellarap detneiro era yehT
.diobmohr era yeht

,submohr detneiro na si e_{12} tcejbo lacitamehtam lanoisnemid-owt ehT
.noitatneiro esiwkcolc-itna na sessessop hcihw

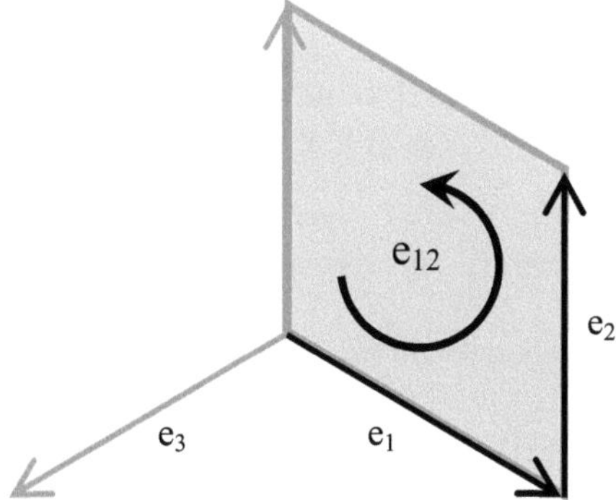

,submohr detneiro na si e$_{21}$ tcejbo lacitamehtam lanoisnemid-owt eht dnA
.noitatneiro esiwkcolc ,desoppo na sessessop hcihw

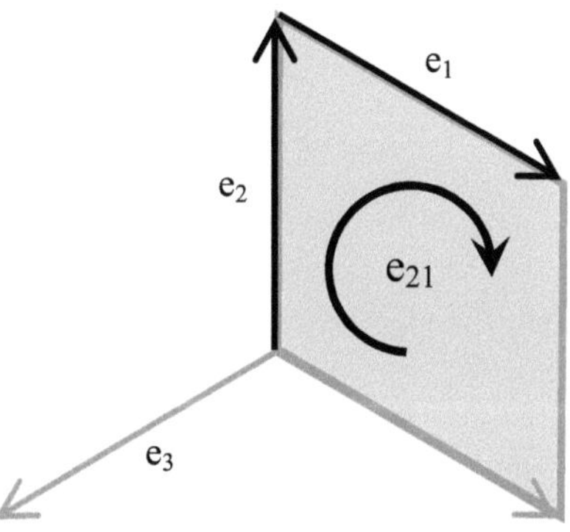

.eromyna seitilauq lairotcev yna evah ton od e$_{21}$ dna e$_{12}$
,sesubmohr detneiro ,lanoisnemid-owt era yehT
.meht edisni peed srebmun evitagen fo yretsym eht edih hcihw
.esruoc fo evitisop era srebmun sehsifrats tnegilletni roF

,evitisop era srebmun evitageN
owt eht fo mus eht sa deterpretni eb ot evah yeht esuaceb
.sesubmohr evitisop owt eht fo mus a sa suht ,e$_{21}$ dna e$_{12}$ secirtam evitisop

.mus siht ot lacitnedi si (omh) eno sunim namuh ehT

$$e_{12} + e_{21} = \begin{pmatrix} 0 & 0 & 1 \\ 1 & 0 & 0 \\ 0 & 1 & 0 \end{pmatrix} + \begin{pmatrix} 0 & 1 & 0 \\ 0 & 0 & 1 \\ 1 & 0 & 0 \end{pmatrix} = \begin{pmatrix} 0 & 1 & 1 \\ 1 & 0 & 1 \\ 1 & 1 & 0 \end{pmatrix} = omh = (-1)$$

.tsixe ton od srebmun evitageN
.stsixe snoitatneiro tnereffid htiw sesubmohr owt fo mus eht ylnO

.[4] ,[3] aes cariD eht fo ruoivaheb eht ot ralimis si sihT

.aes cariD eht fo etats orez deipucco yletelpmoc eht ni seloh era selcitrap-itnA

sesubmohr fo mus eht $\begin{pmatrix} 1 & 1 & 1 \\ 1 & 1 & 1 \\ 1 & 1 & 1 \end{pmatrix} = 0$ etats orez deipucco yletelpmoc eht ni dnA

seno eht fo snoitisop eht ta sorez fo seloh sah $\begin{pmatrix} 0 & 1 & 1 \\ 1 & 0 & 1 \\ 1 & 1 & 0 \end{pmatrix} = e_{21} + e_{12}$

$\cdot \begin{pmatrix} 1 & 0 & 0 \\ 0 & 1 & 0 \\ 0 & 0 & 1 \end{pmatrix} = 1$ xirtam ytitnedi eht fo

.srebmun evitagen fo yretsym citsym eht si sihT
(.[5] eltitbus wen a tog sah [4] koob eht eroferht dnA)

:deunitnoc eb nac noitatupmoc ruo yretsym siht htiW

$$(3 e_1 + e_2 + 2 e_3) e_1 = 3 e_1{}^2 + e_2 e_1 + 2 e_3 e_1 = 3 + e_{21} + 2 e_{12}$$

$$= 2 + e_{12} + \underbrace{1 + e_{12} + e_{21}}_{0} = 2 + e_{12}$$

fo noitacilpitlum eht ot lacitnedi si tluser sihT

$$(2 e_1 + e_3) e_1 = 2 e_1{}^2 + e_3 e_1 = 2 + e_{12}$$

,demrof si aera detneiro lanoisnemid-owt a niagA
margolellarap detneiro na fo mrof ni emit siht
.egap gniwollof eht no nwohs si hcihw

margolellarap siht fo α elgna roiretni ehT
.tcudorp renni eht fo pleh eht htiw dnuof eb nac

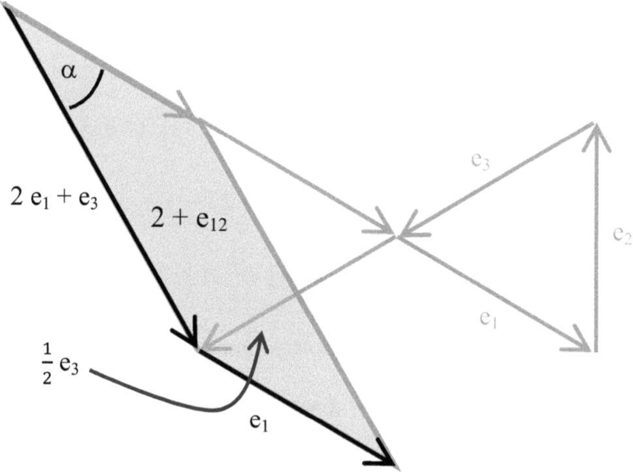

The definition of the inner product of two vectors **a** and **b** is:

$$\mathbf{a} \bullet \mathbf{b} = \frac{1}{2}(\mathbf{a}\,\mathbf{b} + \mathbf{b}\,\mathbf{a})$$

With $\mathbf{a} = 2\,e_1 + e_3$ and $\mathbf{b} = e_1$ we will get:

$$\mathbf{a} \bullet \mathbf{b} = (2\,e_1 + e_3) \bullet e_1 = \frac{1}{2}\left((2\,e_1 + e_3)\,e_1 + e_1\,((2\,e_1 + e_3))\right)$$

$$= \frac{1}{2}(2 + e_{12} + 2 + e_{21}) = \frac{1}{2}(3 + 1 + e_{12} + e_{21}) = \frac{1}{2}(3 + 0) = \frac{3}{2}$$

As the lengths of the vectors are

$$a = |\mathbf{a}| = |2\,e_1 + e_3| = \sqrt{(2\,e_1 + e_3)^2} = \sqrt{3} \approx 1.731 \quad \text{and} \quad b = |\mathbf{b}| = |e_1| = 1$$

the interior angle α will be:

$$\cos\alpha = \frac{\mathbf{a} \bullet \mathbf{b}}{|\mathbf{a}|\,|\mathbf{b}|} = \frac{1}{2}\sqrt{3} \approx 0.8660 \quad \Rightarrow \quad \alpha = 30°$$

And the magnitude of the area $A = |\mathbf{A}|$ of this parallelogram can be found with the help of the outer product. The definition

:si sehsifrats tnegilletni yb nevig **b** dna **a** srotcev owt fo tcudorp retuo eht fo

$$\mathbf{a} \wedge \mathbf{b} = \frac{1}{2}\left(\mathbf{a}\,\mathbf{b} + (e_{12} + e_{21})\,\mathbf{b}\,\mathbf{a}\right)$$

:ni stluser sihT

$$\mathbf{a} \wedge \mathbf{b} = (2\,e_1 + e_3) \wedge e_1 = \frac{1}{2}\left((2\,e_1 + e_3)\,e_1 + (e_{12} + e_{21})\,e_1\,((2\,e_1 + e_3))\right)$$

$$= \frac{1}{2}(2 + e_{12} + 2\,e_{12} + 1 + 2\,e_{21} + e_{12})$$

$$= \frac{1}{2}(3 + 4\,e_{12} + 2\,e_{21}) = \frac{1}{2}(1 + 2\,e_{12}) = \frac{1}{2} + e_{12}$$

noitisopmoced lacinonac eht fo pleh eht htiW

$$\mathbf{a}\,\mathbf{b} = \mathbf{a} \bullet \mathbf{b} + \mathbf{a} \wedge \mathbf{b}$$

:edam eb nac kcehc a

$$\mathbf{a} \bullet \mathbf{b} + \mathbf{a} \wedge \mathbf{b} = \frac{3}{2} + \frac{1}{2} + e_{12} = 2 + e_{12} = (2\,e_1 + e_3)\,e_1 = \mathbf{a}\,\mathbf{b} \quad \Rightarrow \quad \text{o.k.}$$

aera eht fo edutingam eht gnitteg rof noitauqe ehT

$$A = |\mathbf{A}| = |\mathbf{a} \wedge \mathbf{b}| = \sqrt{(\mathbf{a} \wedge \mathbf{b})(\mathbf{b} \wedge \mathbf{a})}$$

:ni stluser neht nevig era **b** dna **a** srotcev ekilecaps erup nehw

$$A = \left|\frac{1}{2} + e_{12}\right| = \sqrt{\left(\frac{1}{2} + e_{12}\right)\left(\frac{1}{2} + e_{21}\right)} = \sqrt{\frac{1}{4} + \frac{1}{2}e_{21} + \frac{1}{2}e_{12} + 1} = \frac{1}{2}\sqrt{3} \approx 0.8660$$

.dekcehc eb nac niaga sihT

,rehto hcae ot ralucidneprep era srotcev owT
.sraeppasid dna orez semoceb tcudorp renni rieht fi

esuaceb ,$2\,e_1 + e_3 = $ **a** rotcev edis eht ot ralucidneprep si e_3 rotcev tinu ehT

$$\mathbf{a} \bullet e_3 = (2\,e_1 + e_3) \bullet e_3 = \frac{1}{2}\left((2\,e_1 + e_3)\,e_3 + e_3\,((2\,e_1 + e_3))\right)$$

$$= \frac{1}{2}(2\,e_{21} + 1 + 2\,e_{12} + 1) = \frac{1}{2}(2 + 2\,e_{12} + 2\,e_{21}) = 0$$

$0.5\,e_3$ rotcev thgieh eht fo noitcerid eht otni gnitniop si e_3 suhT
.(egap suoiverp eht no erugif ees)

.1 tinu esab eht fo flah fo htgnel a evah tsum thgieh eht oS

:eb ot sah aera eht fo edutingam eht dnA

$$A = a\,h = \left|\,\mathbf{a}\,\right|\ \left|\ \tfrac{1}{2}\,e_3\ \right| = \sqrt{3}\cdot\tfrac{1}{2} \approx 0.8660 \qquad \Rightarrow \qquad \text{o.k.}$$

.ytreporp cirtemoeg tnatropmi rehtona si ereht ,yaw eht yB

,rehto hcae ot lellarap era srotcev owT
.sraeppasid dna orez semoceb tcudorp retuo rieht fi

sehsifrats tnegilletni fo arbegla eht fo sthgilhgih cirtemoeG
.snoitator dna snoitcelfer era

yllacitamehtam delledom era snoitator dna snoitcelfer esehT
.stcudorp hciwdnas yb

,n rotcev tinu fo noitcerid eht otni stniop hcihw ,sixa na ni detcelfer si r rotcev A
,thgir eht morf deilpitlum-tsop dna tfel eht morf deilpitlum-erp si n nehw
:hciwdnas lacitamehtam siht fo elddim eht ni dezeeuqs si r rotcev elihw

$$\mathbf{r_{ref}} = \mathbf{n}\ \mathbf{r}\ \mathbf{n}$$

.noitator a secudorp tcudorp hciwdnas elbuod a dnA

$$\mathbf{r_{rot}} = \mathbf{m}\ \mathbf{r_{ref}}\ \mathbf{m} = \mathbf{m}\ \mathbf{n}\ \mathbf{r}\ \mathbf{n}\ \mathbf{m}$$

,$\mathbf{r_{rot}}$ rotcev eht otni demrofsnart neht si r rotcev ehT
sexa owt eht neewteb elgna eht sa gib sa eciwt si noitator fo elgna eht elihw
.m dna n fo snoitcerid eht otni gnitniop era hcihw ,noitcelfer fo

$5\ e_1 + 5\ e_3 = r$ elopgalf eht ta dexif si hcihw galf a elpmaxe roF
detator eb lliw
.(egap gniwollof eht no erugif ees)

:snoitcelfer owt gniwollof eht yb detneserper si noitator sihT

sixa na ni noitcelfer a evah ew tsriF
.$e_1 = \mathbf{n}$ noitcelfer fo rotcev tinu eht fo noitcerid eht otni stniop hcihw

noitcerid eht otni stniop hcihw sixa na ni noitcelfer dnoces a evah ew dnA
$\sqrt{\tfrac{2}{3}}\ e_1 + \dfrac{1+\sqrt{3}}{\sqrt{6}}\ e_2 = \dfrac{1}{\sqrt{6}}\left(2\ e_1 + (1+\sqrt{3})\ e_2\right) = \mathbf{m}$ noitcelfer fo rotcev tinu eht fo $\cdot$

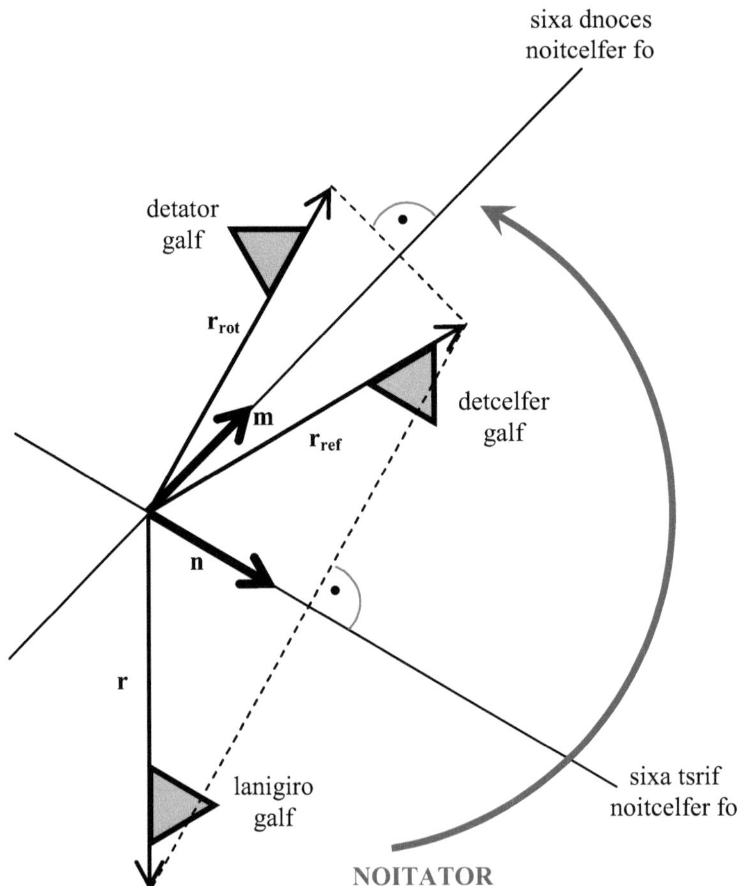

:ot gnidrocca demrofsnart neht si **r** elopgalf eht fo rotcev ehT

$$\mathbf{r_{ref}} = \mathbf{n}\,\mathbf{r}\,\mathbf{n} = e_1\,(5\,e_1 + 5\,e_3)\,e_1 = 5\,e_1{}^2\,e_1 + 5\,e_1e_3e_1 = 5\,e_1 + 5\,e_2$$

,noitcerid evitisop a otni stniop suht elopgalf detcelfer ehT
:e_3 rotcev tinu eht fo noitcerid eht ot desoppo yltcerid si hcihw

$$omh\ e_3 = (e_{12} + e_{21})\,e_3 = e_1e_2e_3 + e_2e_1e_3 = e_1 + e_2$$

,noitcelfer tsrif siht ta detcelfer osla si galf eht sA
.(erugif ees) elopgalf eht fo edis desoppo eht no si galf detcelfer eht fo noitisop eht

elopgalf eht fo edis gnorw eht ot derorrim neeb sah galf ehT
.noitcelfer dnoces a yb edis tcerroc eht ot kcab derorrim eb tsum dna

:si noitcelfer dnoces siht fo tcudorp hciwdnas ehT

$$\mathbf{r_{rot}} = \mathbf{m}\,\mathbf{r_{ref}}\quad \mathbf{m} = \frac{1}{\sqrt{6}}\,(2\,e_1 + (1+\sqrt{3})\,e_2)\ (5\,e_1+5\,e_2)\ \frac{1}{\sqrt{6}}\,(2\,e_1+(1+\sqrt{3})\,e_2)$$

$$= \frac{5}{6}\,(3+\sqrt{3}+2\,e_{12}+e_{21}+\sqrt{3}\,e_{21})\ (2\,e_1+(1+\sqrt{3})\,e_2)$$

$$= \frac{5}{6}\,(2+\sqrt{3}+e_{12}+\sqrt{3}\,e_{21})\ (2\,e_1+(1+\sqrt{3})\,e_2)$$

$$= \frac{5}{6}\,((5+3\sqrt{3})\,e_1+(5+5\sqrt{3})\,e_2+(5+\sqrt{3})\,e_3)$$

$$= \frac{5}{6}\,(2\sqrt{3}\,e_1+4\sqrt{3}\,e_2)\ =\ \frac{5}{3}\,(\sqrt{3}\,e_1+2\sqrt{3}\,e_2)$$

$$= \frac{5}{\sqrt{3}}\,(e_1+2\,e_2)$$

:serauqs fo kcehc trohS

$$\mathbf{r}^2 = (5\,e_1+5\,e_3)^2 = 50+25\,e_{12}+25\,e_{21} = 25$$
$$\mathbf{r_{ref}}^{\,2} = (5\,e_1+5\,e_2)^2 = 50+25\,e_{12}+25\,e_{21} = 25$$
$$\mathbf{r_{rot}}^{\,2} = \frac{25}{3}\,(e_1+2\,e_2)^2 = \frac{25}{3}\,(5+2\,e_{12}+2\,e_{21}) = \frac{25}{3}\,(3+0) = 25$$

$$\left. \right\} \quad \mathbf{r}^2 = \mathbf{r_{ref}}^{\,2} = \mathbf{r_{rot}}^{\,2} \quad \Rightarrow \text{o.k.}$$

:selgna fo kcehC

$$\cos\alpha = \hat{\mathbf{r}}\bullet\hat{\mathbf{r}}_{rot} = \frac{1}{2}\,(\hat{\mathbf{r}}\,\hat{\mathbf{r}}_{rot}+\hat{\mathbf{r}}_{rot}\,\hat{\mathbf{r}})$$

$$= \frac{1}{2\cdot25}\,\Big(5\,e_1+5\,e_3\Big)\frac{5}{\sqrt{3}}\,(e_1+2\,e_2)+\frac{5}{\sqrt{3}}\,(e_1+2\,e_2)\,(5\,e_1+5\,e_3)\Big)$$

$$= \frac{1}{2\sqrt{3}}\,((e_1+e_3)\,(e_1+2\,e_2)+(e_1+2\,e_2)\,(e_1+e_3))$$

$$= \frac{1}{2\sqrt{3}}\,(1+3\,e_{12}+2\,e_{21}+1+2\,e_{12}+3\,e_{21})$$

$$= \frac{1}{2}\,\sqrt{3}\,(e_{12}+e_{21}) \approx 0.8660\,(e_{12}+e_{21}) \qquad \Rightarrow \qquad \alpha = 150°$$

⇐   .150° fo edutingam a sah noitator fo elgna ehT

(.[2] koob egnaro eht daer esaelp ,siht eveileb ton od uoy fi dnA)

$$\cos \beta = \mathbf{n} \bullet \mathbf{m} = \tfrac{1}{2}(\mathbf{n\,m} + \mathbf{m\,n})$$

$$= \tfrac{1}{2}\left(e_1 \; \tfrac{1}{\sqrt{6}}\,(2\,e_1 + (1+\sqrt{3})\,e_2) + \tfrac{1}{\sqrt{6}}\,(2\,e_1 + (1+\sqrt{3})\,e_2)\,e_1\right)$$

$$= \tfrac{1}{2\sqrt{6}}\,(2 + e_{12} + \sqrt{3}\,e_{12} + 2 + e_{21} + \sqrt{3}\,e_{21})$$

$$= \tfrac{1}{2\sqrt{6}}\,(3 + \sqrt{3}\,e_{12} + \sqrt{3}\,e_{21}) = \tfrac{1}{2\sqrt{2}}\,(\sqrt{3} + e_{12} + e_{21})$$

$$= \tfrac{1}{\sqrt{2}+\sqrt{6}} \approx 0.2588 \qquad \Rightarrow \qquad \beta = 75°$$

,evitisop si siht llA

.ereh $\sqrt{3} - 1 = \dfrac{2}{1+\sqrt{3}}$ noitauqe namuh dam eht desu evah ew fi neve

:uoy llet lliw eh dna ,rehposoliph hsifrats lanosrep ruoy ksA

$$\sqrt{3} + e_{12} + e_{21} = (\sqrt{3} + e_{12} + e_{21}) \cdot \frac{\sqrt{3}+1}{\sqrt{3}+1} = \frac{3+e_{12}+e_{21}+\sqrt{3}\,(1+e_{12}+e_{21})}{\sqrt{3}+1} = \frac{2}{1+\sqrt{3}}$$

.sngis evitagen tuohtiw yletelpmoc

$\Leftarrow$. 75° si noitcelfer fo sexa owt eht neewteb elgna ehT

:([2] koob egnaro eht ees) noisulcnoC

.noitcelfer fo sexa eht neewteb elgna eht sa gib sa eciwt si noitator fo elgna ehT

$$\alpha = 2\,\beta$$

.noititeper eht dehsinif evah won eW

.snoitautis laitaps ylerup tuoba si retpahc siht ni dias neeb sah hcihw gnihtyrevE
.ereh deredisnoc neeb ton evah seitimalac laropmet ro detaler-emiT
.sretpahc txen eht ni emoc ot gniog si sihT

!gnihtyreve egnahc lliw emiT

thgiL fo noitnevnI ehT 2

,og dna emoc lliw llits sega ,oga snoeA
,retaw htiw derevoc yletelpmoc dna yawa raf ,raf tenalp ylenol a no
,dediced tsitneics hsifrats dethgis-raf dna ,esiw ,yhs ,gnuoy a
.stsixe emit taht

,emit fo rotcev tinu a dedeen yltnegru eh oS
e_2 rotcev tinu eht taht dediced eh dna
.rotcev ekilemit elbatius dna ecin a eb dluohs

.tnoip gnitrats ruo si sihT

emitecaps lanoisnemid-owt fo snoitcerid eerht eht fo enO
.γ_1 rotcev tinu ekilemit siht fo noitcerid eht otni gnitniop si

$$\gamma_1 = e_2 = \begin{pmatrix} 0 & 0 & 1 \\ 0 & 1 & 0 \\ 1 & 0 & 0 \end{pmatrix}$$

,taht tuoba gnikniht no tnew tsitneics hsifrats esiw ,trams ,gnuoy eht dnA
.stsixe ecaps taht ,dediced eh dna

,ecaps fo rotcev tinu a dedeen osla eh oS
$(1 + 2\,e_{12})$ rotcev eht taht dediced eh dna
.rotcev ekilecaps elbatius dna ecin a eb dluohs

γ_2 rotcev tinu a ti edam tsitneics hsifrats esiw gnuoy ehT
:nwod etorw dna

$$\gamma_2 = \frac{1}{\sqrt{3}}\,(1 + 2\,e_{12}) = \frac{1}{\sqrt{3}}\begin{pmatrix} 1 & 0 & 2 \\ 2 & 1 & 0 \\ 0 & 2 & 1 \end{pmatrix}$$

,hsifrats a saw tsitneics hsifrats eht tuB
.yrtemmys dlofeerht devol ohw

,srotcev tinu eerht dnif ot dah eh suhT
snoitcerid tnereffid eerht otni gnitniop
.emitecaps lanoisnemid-owt detnevni ylwen sih fo

sdoow retawrednu eht hguorht deklaw gnivah retfa tsuj ,gnineve eno dnA
,hcum os devol eh hcihw ,sllih retawrednu gogaM goG eht fo

,shtnom rof gnihton dias gnivah dna eert eagla errazib a debmilc gnivah retfa
emoh denruter tsitneics hsifrats yhs gnuoy eht
,γ_0 rotcev tinu driht gniwollof eht nwod etorw dna
:rotcev tinu ekilecaps a ron rotcev tinu ekilemit a rehtien saw hcihw

$$\gamma_0 = e_1 + e_3 + \frac{1}{\sqrt{3}} + \frac{2}{\sqrt{3}}\, e_{21} = \frac{1}{\sqrt{3}}\begin{pmatrix} 1+\sqrt{3} & 2+\sqrt{3} & 0 \\ \sqrt{3} & 1 & 2+\sqrt{3} \\ 2 & \sqrt{3} & 1+\sqrt{3} \end{pmatrix}$$

.thgil fo noitnevni eht neeb sah sihT
.thgil fo noitcerid eht otni stniop emitecaps lanoisnemid-owt fo noitcerid driht ehT

.emit fo trap ton si thgiL .ecaps fo trap ton si thgiL
tnenopmoc elbativeni ,elbapacseni ,yratnemele na ,emitecaps fo trap si thgiL
.emit dna ecaps gnitarapes ,dlrow ruo fo

tsitneics hsifrats eht neht dnA
.metsys etanidrooc hsifrats gniwollof eht dehcteks
.ytivitaler hsifrats fo metsys etanidrooc eht si tI

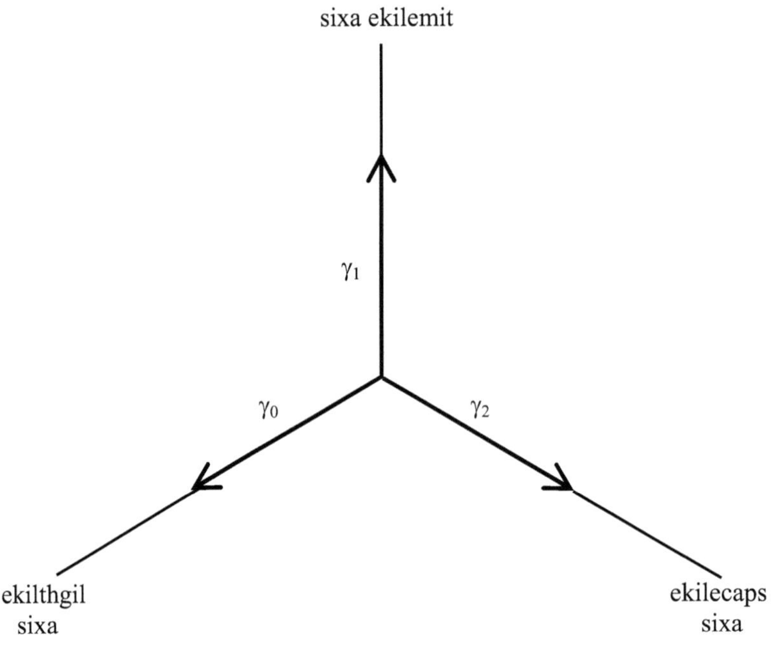

23

:deksa tsitneics hsifrats eht fo seugaelloc eht dnA
?ekilemit γ_1 llac uoy od yhW
?ekilecaps γ_2 llac uoy od yhW
?ekilthgil γ_0 llac uoy od yhW
?taht rof nosaer lacitamehtam eht si tahW

:dednopser tsitneics hsifrats eht dnA

,rotcev ekilemit a eb tsum rotcev A
,eno fo elpitlum a si erauqs sti fi
.xirtam ytitnedi eht fo elpitlum a

,emit fo rotcev tinu eht gnirauqs detrats eh dnA
.klaF fo emehcs eht wenk eh sa

$$\gamma_1^{\,2} \quad \begin{array}{ccc} 0 & 0 & 1 \\ 0 & 1 & 0 \\ 1 & 0 & 0 \end{array}$$

$$\begin{array}{ccc|ccc} 0 & 0 & 1 & 1 & 0 & 0 \\ 0 & 1 & 0 & 0 & 1 & 0 \\ 1 & 0 & 0 & 0 & 0 & 1 \end{array} \quad \Rightarrow \quad \gamma_1^{\,2} = 1$$

:nialpxe ot no tnew tsitneics hsifrats eht dnA

,($e_{12} + e_{21}$) fo elpitlum a si erauqs sti fi ,rotcev ekilecaps a eb tsum rotcev A
xirtam deipucco yletelpmoc eht fo elpitlum a
.seno eht fo snoitisop eht ta seloh s'cariD htiw

.ecaps fo rotcev tinu eht gnirauqs detrats eh dnA

,snoitcarf diova oT
. $\sqrt{3}\,\gamma_2$ xirtam eht derauqs ylizal yrev tsitneics hsifrats eht

.noitatneserper dradnats eht otni tluser eht demrofsnart eh esruoc fo dnA

$(\sqrt{3}\,\gamma_2)^2 = 3\,\gamma_2^2$			1	0	2
			2	1	0
			0	2	1
1	0	2	1	4	4
2	1	0	4	1	4
0	2	1	4	4	1

:noitatneserper dradnats eht otni noitamrofsnarT

$$\Rightarrow \quad 3\,\gamma_2^2 = \begin{pmatrix} 1 & 4 & 4 \\ 4 & 1 & 4 \\ 4 & 4 & 1 \end{pmatrix} = \begin{pmatrix} 0 & 3 & 3 \\ 3 & 0 & 3 \\ 3 & 3 & 0 \end{pmatrix} = 3\,(e_{12} + e_{21})$$

$$\Rightarrow \quad \gamma_2^2 = \begin{pmatrix} 0 & 1 & 1 \\ 1 & 0 & 1 \\ 1 & 1 & 0 \end{pmatrix} = e_{12} + e_{21}$$

,rotcev tinu tsal sih htiw seugaelloc sih delffab tsitneics hsifrats eht yllanif dnA
:rotcev tinu lamron a eb ot tnaw t'ndid spahrep hcihw

,rotcev ekilthgil a eb tsum rotcev A
,xirtam llun eht fo elpitlum a suht dna orez fo elpitlum a si erauqs sti fi
.flesti llun tsuj gnieb

$(\sqrt{3}\,\gamma_0)^2 = 3\,\gamma_0^2$			$1 + \sqrt{3}$	$2 + \sqrt{3}$	0
			$\sqrt{3}$	1	$2 + \sqrt{3}$
			2	$\sqrt{3}$	$1 + \sqrt{3}$
$1 + \sqrt{3}$	$2 + \sqrt{3}$	0	$7 + 4\sqrt{3}$	$7 + 4\sqrt{3}$	$7 + 4\sqrt{3}$
$\sqrt{3}$	1	$2 + \sqrt{3}$	$7 + 4\sqrt{3}$	$7 + 4\sqrt{3}$	$7 + 4\sqrt{3}$
2	$\sqrt{3}$	$1 + \sqrt{3}$	$7 + 4\sqrt{3}$	$7 + 4\sqrt{3}$	$7 + 4\sqrt{3}$

:noitatneserper dradnats eht otni noitamrofsnarT

$$\Rightarrow \qquad 3\,\gamma_0^{\ 2} = \begin{pmatrix} 7+\sqrt{48} & 7+\sqrt{48} & 7+\sqrt{48} \\ 7+\sqrt{48} & 7+\sqrt{48} & 7+\sqrt{48} \\ 7+\sqrt{48} & 7+\sqrt{48} & 7+\sqrt{48} \end{pmatrix} = \begin{pmatrix} 0 & 0 & 0 \\ 0 & 0 & 0 \\ 0 & 0 & 0 \end{pmatrix} = 0$$

$$\Rightarrow \qquad \gamma_0^{\ 2} = \begin{pmatrix} 0 & 0 & 0 \\ 0 & 0 & 0 \\ 0 & 0 & 0 \end{pmatrix} = 0$$

.stsixe orez siht esuaceb ,stsixe thgiL

emitecaps lanoisnemid-owt a fo rotcev yrevE
noitanibmoc raenil a sa detneserper eb suht nac

emit dna ecaps	•	fo
thgil dna emit	•	fo ro
:ecaps dna thgil	•	fo ro

$$\mathbf{r} = c\,t\,\gamma_1 + x\,\gamma_2 + y\,\gamma_0$$

.tsael ta orez eb lliw y dna ,x ,c t setanidrooc eerht eht fo eno dnA
.deedni ,emitecaps lanoisnemid-owt a si tI

.deilpitlum eb nac secirtaM
.deilpitlum eb nac srotcev tinu suhT
.srotcev tinu eseht fo seitreporp larutcurts cisab eht slaever sihT

.arbegla cariD dellac si erutcurts siht smetsys lacitamehtam namuh nI
snoitauqe gniwollof eht eroferehT
.sehsifrats fo arbegla cariD eht ebircsed

sehsifrats fo arbegla cariD siht fo tniop gnitrats ehT
srotcev tinu sa γ_2 dna γ_1 srotcev eerht eht yb nevig si
… rotcev tinu eb-dluow sa γ_0 dna

$$\gamma_1 = e_2 = \begin{pmatrix} 0 & 0 & 1 \\ 0 & 1 & 0 \\ 1 & 0 & 0 \end{pmatrix} \qquad \gamma_2 = \frac{1}{\sqrt{3}}(1 + 2\,e_{12}) = \frac{1}{\sqrt{3}}\begin{pmatrix} 1 & 0 & 2 \\ 2 & 1 & 0 \\ 0 & 2 & 1 \end{pmatrix}$$

$$\gamma_0 = e_1 + e_3 + \frac{1}{\sqrt{3}} + \frac{2}{\sqrt{3}}\,e_{21} = \frac{1}{\sqrt{3}}\begin{pmatrix} 1+\sqrt{3} & 2+\sqrt{3} & 0 \\ \sqrt{3} & 1 & 2+\sqrt{3} \\ 2 & \sqrt{3} & 1+\sqrt{3} \end{pmatrix}$$

:snoitidnoc noitazilamron nwonk ydaerla eht htiw rehtegot …

$$\gamma_1{}^2 = \begin{pmatrix} 1 & 0 & 0 \\ 0 & 1 & 0 \\ 0 & 0 & 1 \end{pmatrix} = 1 \qquad \gamma_2{}^2 = \begin{pmatrix} 0 & 1 & 1 \\ 1 & 0 & 1 \\ 1 & 1 & 0 \end{pmatrix} = e_{12} + e_{21} \qquad \gamma_0{}^2 = \begin{pmatrix} 0 & 0 & 0 \\ 0 & 0 & 0 \\ 0 & 0 & 0 \end{pmatrix} = 0$$

:ylpitlum ew woN

$$\gamma_1\,\gamma_2 = e_2\,\frac{1}{\sqrt{3}}(1 + 2\,e_{12}) = \frac{1}{\sqrt{3}}\,e_2 + \frac{2}{\sqrt{3}}\,e_3$$

$$\gamma_2\,\gamma_1 = \frac{1}{\sqrt{3}}(1 + 2\,e_{12})\,e_2 = \frac{1}{\sqrt{3}}\,e_2 + \frac{2}{\sqrt{3}}\,e_1$$

:noisulcnoc tsriF

$$\Rightarrow \qquad \gamma_1\,\gamma_2 + \gamma_2\,\gamma_1 = \frac{2}{\sqrt{3}}\,e_1 + \frac{2}{\sqrt{3}}\,e_2 + \frac{2}{\sqrt{3}}\,e_3 = \begin{pmatrix} 0 & 0 & 0 \\ 0 & 0 & 0 \\ 0 & 0 & 0 \end{pmatrix} = 0$$

:snoitacilpitlum fo nosirapmoc dnoceS

$$\gamma_2\,\gamma_0 = \frac{1}{3}\,(1 + 2\,e_{12})\,(\sqrt{3}\,e_1 + \sqrt{3}\,e_3 + 1 + 2\,e_{21})$$

$$= \frac{1}{3}\,(\sqrt{3}\,e_1 + 2\sqrt{3}\,e_2 + 3\sqrt{3}\,e_3 + 5 + 2\,e_{12} + 2\,e_{21}) = \frac{1}{\sqrt{3}}\,e_2 + \frac{2}{\sqrt{3}}\,e_3 + 1$$

$$\gamma_0\,\gamma_2 = \frac{1}{3}\,(\sqrt{3}\,e_1 + \sqrt{3}\,e_3 + 1 + 2\,e_{21})\,(1 + 2\,e_{12})$$

$$= \frac{1}{3}\,(3\sqrt{3}\,e_1 + 2\sqrt{3}\,e_2 + \sqrt{3}\,e_3 + 5 + 2\,e_{12} + 2\,e_{21}) = \frac{2}{\sqrt{3}}\,e_1 + \frac{1}{\sqrt{3}}\,e_2 + 1$$

:noisulcnoc dnoceS

$$\Rightarrow \quad \gamma_2\,\gamma_0 + \gamma_0\,\gamma_2 = \frac{2}{\sqrt{3}}\,e_1 + \frac{2}{\sqrt{3}}\,e_2 + \frac{2}{\sqrt{3}}\,e_3 + 2 = \begin{pmatrix} 2 & 0 & 0 \\ 0 & 2 & 0 \\ 0 & 0 & 2 \end{pmatrix} = 2$$

:snoitacilpitlum fo nosirapmoc drihT

$$\gamma_0\,\gamma_1 = \frac{1}{\sqrt{3}}\,(\sqrt{3}\,e_1 + \sqrt{3}\,e_3 + 1 + 2\,e_{21})\,e_2 = e_{12} + e_{21} + \frac{1}{\sqrt{3}}\,e_2 + \frac{2}{\sqrt{3}}\,e_3$$

$$\gamma_1\,\gamma_0 = \frac{1}{\sqrt{3}}\,e_2\,(\sqrt{3}\,e_1 + \sqrt{3}\,e_3 + 1 + 2\,e_{21}) = e_{12} + e_{21} + \frac{2}{\sqrt{3}}\,e_1 + \frac{1}{\sqrt{3}}\,e_2$$

:noisulcnoc drihT

$$\Rightarrow \quad \gamma_1\,\gamma_0 + \gamma_0\,\gamma_1 = \frac{2}{\sqrt{3}}\,e_1 + \frac{2}{\sqrt{3}}\,e_2 + \frac{2}{\sqrt{3}}\,e_3 + 2\,e_{12} + 2\,e_{21} = \begin{pmatrix} 0 & 2 & 2 \\ 2 & 0 & 2 \\ 2 & 2 & 0 \end{pmatrix} = 2\,(e_{12} + e_{21})$$

egap gniwollof eht ta yrammus ehT
tcudorp renni eht esirpmoc osla lliw

$$\mathbf{a} \bullet \mathbf{b} = \frac{1}{2}\,(\mathbf{a}\,\mathbf{b} + \mathbf{b}\,\mathbf{a})$$

[6] lufituaeb yllaer ,yllaer kool lliw siht dnA
![7] eurt yllaer ,yllaer dna

:era arbegla cariD hsifrats fo snoitauqe cisab ehT

$$\gamma_0{}^2 = \frac{1}{2}\left(\gamma_1\,\gamma_2 + \gamma_2\,\gamma_1\right) = \gamma_1 \bullet \gamma_2 = \begin{pmatrix} 0 & 0 & 0 \\ 0 & 0 & 0 \\ 0 & 0 & 0 \end{pmatrix} = 0$$

$$\gamma_1{}^2 = \frac{1}{2}\left(\gamma_2\,\gamma_0 + \gamma_0\,\gamma_2\right) = \gamma_2 \bullet \gamma_0 = \begin{pmatrix} 1 & 0 & 0 \\ 0 & 1 & 0 \\ 0 & 0 & 1 \end{pmatrix} = 1$$

$$\gamma_2{}^2 = \frac{1}{2}\left(\gamma_0\,\gamma_1 + \gamma_1\,\gamma_0\right) = \gamma_0 \bullet \gamma_1 = \begin{pmatrix} 0 & 1 & 1 \\ 1 & 0 & 1 \\ 1 & 1 & 0 \end{pmatrix} = e_{12} + e_{21}$$

,og dna emoc lliw llits snoeA
.wolf sega eht sa esirprus lliw snoitauqe cariD eseht tub

tsixE toN oD srebmuN evitageN 4

eert eagla retawrednu sih ni gnittis saw tsitneics hsifrats eht dnA
.srotcev sih tuoba sthguoht ni tsol yllatot saw eh dna

.erutan era yehT .gniwolbdnim era yehT .lufituaeb era yehT

gnihtoN = thgiL + emiT + ecapS

:orez ot pu dda dlrow ruo fo stneutitsnoc eerht eht fo srotcev tinu eerht ehT

$$\gamma_0 + \gamma_1 + \gamma_2 = \frac{1}{\sqrt{3}} \begin{pmatrix} 2+\sqrt{3} & 2+\sqrt{3} & 2+\sqrt{3} \\ 2+\sqrt{3} & 2+\sqrt{3} & 2+\sqrt{3} \\ 2+\sqrt{3} & 2+\sqrt{3} & 2+\sqrt{3} \end{pmatrix} = \begin{pmatrix} 0 & 0 & 0 \\ 0 & 0 & 0 \\ 0 & 0 & 0 \end{pmatrix} = 0$$

traeh lacihposolihp eht setutitsnoc $\gamma_0 + \gamma_1 + \gamma_2 = 0$ mus llun sihT
.dlrow lacisyhp ruo fo

,oot ,serauqs ni gnitaeb si traeh lacihposolihp lacisyhp sihT

$$\gamma_0 + \gamma_1 + \gamma_2 = 0$$

$$\gamma_0{}^2 + \gamma_1{}^2 + \gamma_2{}^2 = \gamma_1{}^2 + \gamma_2{}^2 = 0$$

.thgil tuohtiw gninethgilne ,emit emas eht ta sselthgil dna thgirb

.tsixe ton od srebmun evitageN
.tsixe emit fo serauqs dna ecaps fo serauqs ylnO
.seulav evitagen nwod etirw reven sehsifratS
.srebmun evitagen esu reven sehsifratS

,sraeppa omh ro (− 1) rebmun namuh drusba eht erehwemos revenehW
,noitatiseh tuohtiw dna yletaidemmi ti ecalper lliw sehsifrats
. $\gamma_2{}^2$ ecaps fo erauqs tinu eht yb yltnatsni

,sraeppa rotcev evitagen namuh drusba na revenehw dnA
,noitatiseh tuohtiw dna yletaidemmi ti ecalper lliw sehsifrats
:srotcev owt rehto eht fo noitanibmoc raenil evitisop a yb yltnatsni

$$\gamma_2{}^2\,\gamma_0 = \gamma_1 + \gamma_2 \qquad \gamma_2{}^2\,\gamma_1 = \gamma_2 + \gamma_0 \qquad \gamma_2{}^2\,\gamma_2 = \gamma_0 + \gamma_1$$

snoitaler eerht eseht kcehc nac eW
:snoitatneserper xirtam eht fo pleh eht htiw

$$\gamma_2{}^2\,\gamma_0 = \frac{1}{\sqrt{3}}\begin{pmatrix} 2+\sqrt{3} & 1+\sqrt{3} & 3+2\sqrt{3} \\ 3+\sqrt{3} & 2+2\sqrt{3} & 1+\sqrt{3} \\ 1+2\sqrt{3} & 3+\sqrt{3} & 2+\sqrt{3} \end{pmatrix} = \frac{1}{\sqrt{3}}\begin{pmatrix} 1 & 0 & 2+\sqrt{3} \\ 2 & 1+\sqrt{3} & 0 \\ \sqrt{3} & 2 & 1 \end{pmatrix}$$

$$\gamma_1+\gamma_2 = \frac{1}{\sqrt{3}}\begin{pmatrix} 1 & 0 & 2+\sqrt{3} \\ 2 & 1+\sqrt{3} & 0 \\ \sqrt{3} & 2 & 1 \end{pmatrix} \qquad\Rightarrow\qquad \gamma_2{}^2\,\gamma_0 = \gamma_1+\gamma_2$$

dna

$$\gamma_2{}^2\,\gamma_1 = \begin{pmatrix} 1 & 1 & 0 \\ 1 & 0 & 1 \\ 0 & 1 & 1 \end{pmatrix}$$

$$\Rightarrow\qquad \gamma_2{}^2\,\gamma_1 = \gamma_2+\gamma_0$$

$$\gamma_2+\gamma_0 = \frac{1}{\sqrt{3}}\begin{pmatrix} 2+\sqrt{3} & 2+\sqrt{3} & 2 \\ 2+\sqrt{3} & 2 & 2+\sqrt{3} \\ 2 & 2+\sqrt{3} & 2+\sqrt{3} \end{pmatrix} = \begin{pmatrix} 1 & 1 & 0 \\ 1 & 0 & 1 \\ 0 & 1 & 1 \end{pmatrix}$$

dna

$$\gamma_2{}^2\,\gamma_2 = \frac{1}{\sqrt{3}}\begin{pmatrix} 2 & 3 & 1 \\ 1 & 2 & 3 \\ 3 & 1 & 2 \end{pmatrix} = \frac{1}{\sqrt{3}}\begin{pmatrix} 1 & 2 & 0 \\ 0 & 1 & 2 \\ 2 & 0 & 1 \end{pmatrix}$$

$$\Rightarrow\qquad \gamma_2{}^2\,\gamma_2 = \gamma_0+\gamma_1$$

$$\gamma_0+\gamma_1 = \frac{1}{\sqrt{3}}\begin{pmatrix} 1+\sqrt{3} & 2+\sqrt{3} & \sqrt{3} \\ \sqrt{3} & 1+\sqrt{3} & 2+\sqrt{3} \\ 2+\sqrt{3} & \sqrt{3} & 1+\sqrt{3} \end{pmatrix} = \frac{1}{\sqrt{3}}\begin{pmatrix} 1 & 2 & 0 \\ 0 & 1 & 2 \\ 2 & 0 & 1 \end{pmatrix}$$

cariD hsifrats fo snoitauqe cisab eht htiw yltcerid etaluclac ot elbissop si ti esruoc fO
,seitinrete ecnis siht od snaicitamehtam hsifrats ,wonk ew sa raf sa dnA .arbegla
:si ti .g.e

γ_0 semit	$\gamma_0{}^2+\gamma_1{}^2+\gamma_2{}^2 = \gamma_1{}^2+\gamma_2{}^2 = 1+\gamma_2{}^2 = 0$
$\gamma_1+\gamma_2$ sulp	$\gamma_0+\gamma_2{}^2\,\gamma_0 = 0$
mus llun	$\gamma_0+\gamma_1+\gamma_2+\gamma_2{}^2\,\gamma_0 = \gamma_1+\gamma_2$
	$0+\gamma_2{}^2\,\gamma_0 = \gamma_1+\gamma_2$
$\Rightarrow$	$\gamma_2{}^2\,\gamma_0 = \gamma_1+\gamma_2$

.elpicnirp ni deriuqer ton era snoitatneserper xirtaM
,tcurtsnoc yrailixua edamnam a dna namuh a era yehT
.slatrom yranidro su rof sehcturc lacitamehtam laicifitra

emiT fo terceS ehT 5

.edam era stnemerusaem woN .strats scisyhp woN
.tnemerusaem fo tinu a deen yeht ,gnirusaem era sehsifrats esuaceb dnA

.sretem ni gnihtyreve gnirusaem era sehsifratS

,sretem ni derusaem eb lliw gnihtyreve ,thgil neve ,emit nevE
,naicitamehtam hsifrats tnatropmi na ,ikswokniM nnamreH esuaceb
21 rebmetpeS ta siht deredro sah ,siht dediced sah ylbanosaer yrev
.oga gnol [9] ,[8] engoloC ytic retawrednu hsifrats eht ni

.nietsniE ekil ecaps otni emit dna emit otni ecaps mrofsnart ot detnaw tsuj eH
,krow ot gniog ylno si siht dnA
.stinu lcaitnedi evah seititne htob fi

?smetsys etanidrooc hsifrats ni neppah won lliw tahW

.gnissap si emiT .swolf emiT . gniteelf si emiT

γ_1 rotcev tinu eno evom lliw sehsifrats ,sessap emit retem eno fI
.metsys etanidrooc rieht fo sixa emit eht fo noitcerid eht otni
γ_1 srotcev tinu 300 evom lliw sehsifrats , sessap emit (300 m) fI
.metsys etanidrooc rieht fo sixa emit eht fo noitcerid eht otni

,emit tuoba ksa yltnenamrep ohw ,sdnim suovreN
sehctaw tekcop rieht fo sretniop eht taht ,eciton neht lliw
yltcaxe devom evah lliw

$$\Delta t = \frac{\Delta(ct)}{c} = \frac{300\ m}{3 \cdot 10^8\ m/s} = 1 \cdot 10^{-6}\ s = 1\ \mu s$$

,[8] "ztiwnepperT" a siht llac snaicitamehtam hsifrats dethgis-raF
,nialpxe dna [9] "tiw esacriats" ,sriats no dlot ekoj a
.ni dekcik eb lliw thgil fo noitagaporp eht fo yticolev eht taht

,naem dna yaw siht gniklat era sehsifratS
Δ(ct) ecnereffid etanidrooc eht taht
300 000 km/s = c thgil fo yticolev eht yb dedivid eb tsum
.emit namuh eht teg ot

.terces a sah emit eht tuB
.ssap ylno ton seod emiT

Time does not only fly regularly and constantly in only one direction.
Time is fleeting,
and starfish time is fleeting differently in different directions.

In the following starfish coordinate system
an example of this mysterious behaviour of time is shown.

A first starfish is the observer whose time is flowing constantly
along the timelike coordinate axis of the shown coordinate system.
He observes another, second starfish
who is moving from spacetime point A to a second spacetime point B.

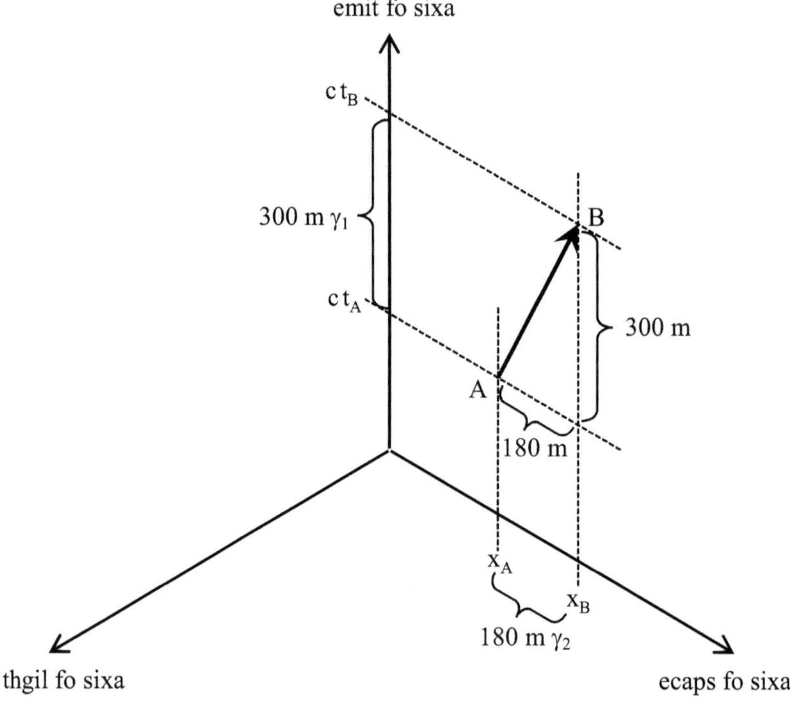

That's all, what is happening in relativity:
An observer observes.

.revresbo eht fo metsys etanidrooc eht ni yb gniog si (300 m) fo emit a ylsuoivbO
,thgir eht ot (180 m) gnivom si hsifrats dnoces eht sessap emit siht elihW
.yawliar noitativel citengam retawrednu kciuq yrev a ni gnittis spahrep

.evisserpmi deedni si yawliar noitativel citengam siht fo yticolev ehT
:si tI

$$v = \frac{\Delta x}{\Delta t} = \frac{180\ \text{m}}{1 \cdot 10^{-6}\ \text{s}} = 1.8 \cdot 10^8\ \frac{\text{m}}{\text{s}} = 180\,000\ \frac{\text{km}}{\text{s}}$$

ro

$$\frac{v}{c} = \frac{\Delta x}{\Delta(ct)} = \frac{180\ \text{m}}{300\ \text{m}} = 0.6 \qquad \Rightarrow \qquad v = 0.6\ c$$

?hsifrats yawliar noitativel dnoces siht rof sessap emit hcum woh tuB

detautis si hsifrats dnoces eht ,sevom yawliar noitativel citengam eht sA
.yawliar noitativel citengam eht fo metsys etanidrooc eht ni

hsifrats dnoces eht neewteb ecnatsid laitaps ehT
,egnahc ton seod tsuj nibac yawliar retawrednu eht dna
.tnatsnoc si ti
,elbatrofmoc yrev era syawliar noitativel citengam retawrednU
.egayov eht gnirud ekahs ton seod hsifrats dnoces eht fo taes eht dna

hsifrats dnoces eht fo noitisop laitaps eht sA
,yawliar retawrednu eht fo metsys etanidrooc eht ni egnahc ton seod
r_{AB} snoitisop emitecaps eht fo egnahc eht fo rotcev nwohs eht
.hsifrats dnoces siht yb derusaem tnenopmoc ekilecaps a evah ton tsum

.hsifrats dnoces eht rof emit fo rotcev erup a eb tsum r_{AB} rotcev ehT
erusaem lliw hsifrats dnoces eht ,emit eht si tI
.yawliar noitativel citengam retawrednu eht edisni

sixa ekilemit eht fo noitcerid eht otni stniop rotcev emit siht eroferehT
,yawliar noitativel citengam retawrednu eht fo metsys etanidrooc eht fo
.egap gniwollof eht fo erugif eht ni nwohs si ti sa tsuj

,detupmoc eb nac rotcev emit siht fo htgnel eht dnA
,erugif tsrif eht ni nevig era rotcev siht fo stnenopmoc htob esuaceb
:revresbo sa hsifrats tsrif eht yb nees

$$r_{AB} = 300\ \text{m}\ \gamma_1 + 180\ \text{m}\ \gamma_2$$

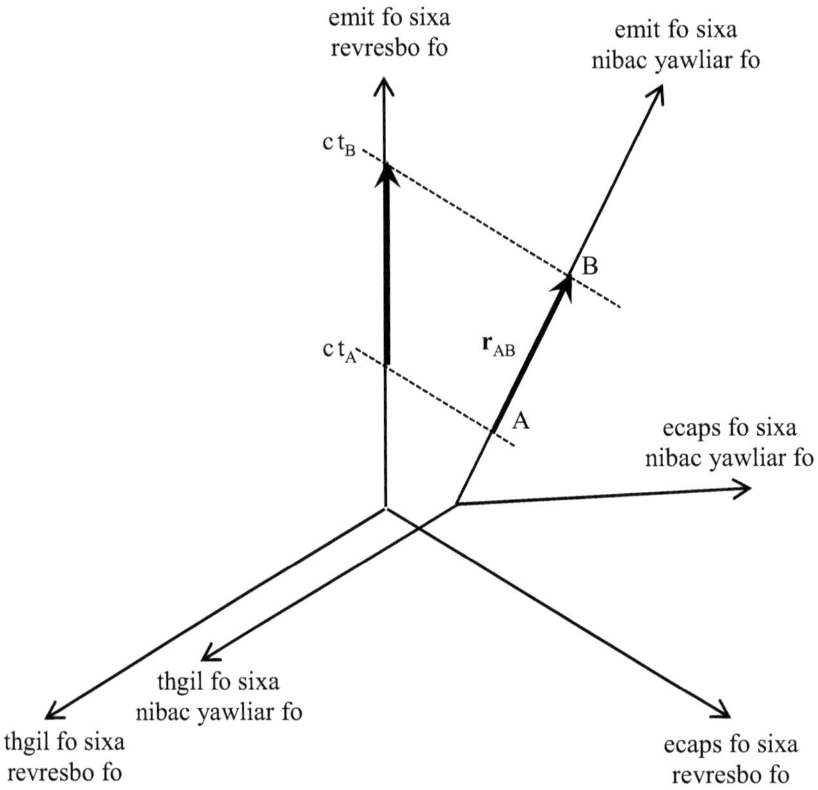

… noitauqe dexaler ,trohs sihT

… ylno emiT = $\mathbf{r}_{AB}$ = 300 m γ_1 + 180 m γ_2 = emiT + ecapS …

… eroc latnem eht sebircsed …
…yretsym etani s'erutan ,aedi cisab eht …
.ytivitaleR laicepS s'nietsniE fo …

tser ta revresbo gnivomnu eht fo metsys etanidrooc tsrif eht nI
.emit dna ecaps fo erutxim a si $\mathbf{r}_{AB}$ rotcev ecnatsid eht
.γ_2 dna γ_1 srotcev tinu eht fo noitanibmoc raenil a si $\mathbf{r}_{AB}$

noitativel citengam retawrednu eht fo metsys etanidrooc gnivom ,dnoces eht nI
.emit fo rotcev erup a si $\mathbf{r}_{AB}$ rotcev ecnatsid eht yawliar

ylno (emirp eno ammag deman) γ_1' rotcev tinu eht fo elpitlum a si $\mathbf{r}_{AB}$
.trap ekilecaps yna tuohtiw

!$\mathbf{r}_{AB}$ rotcev lacitnedi yletelpmoc ,emas eht si siht dnA
.sevitcepsrep tnereffid morf yltnereffid skool ylno tI

:won derauqs eb nac tI

$$\mathbf{r}_{AB}{}^2 = (300 \text{ m } \gamma_1 + 180 \text{ m } \gamma_2)^2$$

$$= 90000 \text{ m}^2\, \gamma_1{}^2 + 54000 \text{ m}^2\, \gamma_1\gamma_2 + 54000 \text{ m}^2\, \gamma_2\gamma_1 + 32400 \text{ m}^2\, \gamma_2{}^2$$

deifilpmis eb nac tluser etaidemretni sihT
:otni arbegla cariD hsifrats fo snoitauqe cisab eht htiw

$$\mathbf{r}_{AB}{}^2 = 90000 \text{ m}^2 \cdot 1 + 32400 \text{ m}^2\, (e_{12} + e_{21}) + 54000 \text{ m}^2\, (\gamma_1\gamma_2 + \gamma_2\gamma_1)$$

$$= 57600 \text{ m}^2 \cdot 1 + 32400 \text{ m}^2 \cdot 1 + 32400 \text{ m}^2\, (e_{12} + e_{21}) + 54000 \text{ m}^2 \cdot 0$$

$$= 57600 \text{ m}^2 \cdot 1 \; + 32400 \text{ m}^2\, (1 + e_{12} + e_{21})$$

$$= 57600 \text{ m}^2 \cdot 1 \; + 32400 \text{ m}^2 \cdot 0$$

$$= 57600 \text{ m}^2$$

$$\Rightarrow r_{AB} = |\mathbf{r}_{AB}| = \sqrt{57600 \text{ m}^2} = 240 \text{ m}$$

yawliar noitativel citengam retawrednu eht ni hsifrats ehT
,(240 m) ylno fo emit a serusaem
.(300 m) fo emit a serusaem hsifrats gnivresbo eht elihw

:sehctaw tekcop esicerp yrev rieht onto kool yeht fi rO

$$\Delta t = \frac{\Delta(ct)}{c} = \frac{300 \text{ m}}{3 \cdot 10^8 \text{ m/s}} = 1 \cdot 10^{-6}\, s = 1\ \mu s$$

na yb derusaem
tser ta revresbo

$$\Delta\tau = \frac{\Delta(c\tau)}{c} = \frac{240 \text{ m}}{3 \cdot 10^8 \text{ m/s}} = 0.8 \cdot 10^{-6}\, s = 0.8\ \mu s$$

yrev eht yb derusaem
hsifrats gnivom tsaf

.emit fo terces eht si sihT

.yltnereffid sessap emiT
.stcejbo ro snosrep gnivom tsaf rof ssap lliw emit sseL

setanidrooc ecaps Δx = v Δt dna emit fo Δt secnereffid eht sA
,v yticolev eht yb detcennoc era
.dnuof eb nac noitauqe latnemadnuf yllacisyhp dna lareneg a

$$\left(\Delta(c\,\tau)\,\gamma_1{}'\right)^2 = \left(\Delta(c\,t)\,\gamma_1 + \Delta x\,\gamma_2\right)^2$$

,sarogahtyP emitecaps dellac osla si noitauqe cisab sihT
elgnairt emitecaps delgna-thgir a fo srotcev edis eht fo serauqs eht esuaceb
.ereh rehto hcae ot deraler era

,segnahc reven hcihw ,tnatsnoc latnemadnuf a si c thgil fo yticolev eht sA
:sa nettirw eb ylevitanretla nac sarogahtyP emitecaps eht

$$c^2\,\Delta\tau^2\,\gamma_1{}'^2 = \left(c\,\Delta t\,\gamma_1 + \Delta x\,\gamma_2\right)^2$$

:ni stluser siht arbegla cariD gniylppa yB

$$\Delta\tau^2\,\gamma_1{}'^2 = \Delta t^2\left(\gamma_1 + \frac{\Delta x}{c\,\Delta t}\,\gamma_2\right)^2 = \Delta t^2\left(\gamma_1 + \frac{v}{c}\,\gamma_2\right)^2$$

$$\Delta\tau^2\,\gamma_1{}'^2 = \Delta t^2\left(\underset{\uparrow}{\gamma_1^2} + \underbrace{\frac{v}{c}\,\gamma_1\gamma_2 + \frac{v}{c}\,\gamma_2\gamma_1}_{} + \frac{v^2}{c^2}\,\gamma_2^2\right) = \Delta t^2\left(1 + \frac{v^2}{c^2}(e_{12}+e_{21})\right)$$

$$1 \qquad\qquad 0$$

$$\Rightarrow \qquad \Delta\tau = \Delta t\,\sqrt{1 + (e_{12}+e_{21})\frac{v^2}{c^2}} = \Delta t\,\sqrt{1 + \gamma_2{}^2\frac{v^2}{c^2}}$$

tnatropmi yrev dna latnemadnuf ,cisab yllaer ,yllaer a si noitauqe sihT
.ytivitaleR laicepS fo noitauqe

.scisyhp ni **noitalid emit** dellac si tceffe suoiretsym sihT

(ffuts rehto ro skcolc ekil) stcejbo ro snosrep fo Δτ secnatsid emiT
sevlesmeht yb derusaem seiticolev hgih htiw gnivom
tser ta srevresbo ro (skcolc) stcejbo yb derusaem Δt secnatsid emit naht retrohs era
.stcejbo gnivom eht hctaw ylno hcihw dna ohw

,sevlesmeht stcejbo ro snosrep gnivom eht yb derusaem era Δτ secnatsid emit eht sA
.**emit reporp** dellac semitemos era secnatsid emit eseht

.suoiretsym si emit ylno ton tuB
.ytivitaleR laicepS ni suoiretsym dna egnarts sevaheb ,oot ,ecapS

:gnicnivnoc si taht rof nosaer lacitamehtam ehT

.sedis eerht evah selgnairT
.edis ekilecaps ,laitaps a si elgnairt emitecaps a fo edis driht ehT
,sthgnel ekilecaps gninrecnoc tceffe suoiretsym a si ereht eroferehT
.oot ,sthgnel reporp tub ,semit reporp redisnoc ot evah ylno ton stsitneics taht hcus

.retal liated erom ni dessucsid eb lliw ecaps fo terces siht dnA

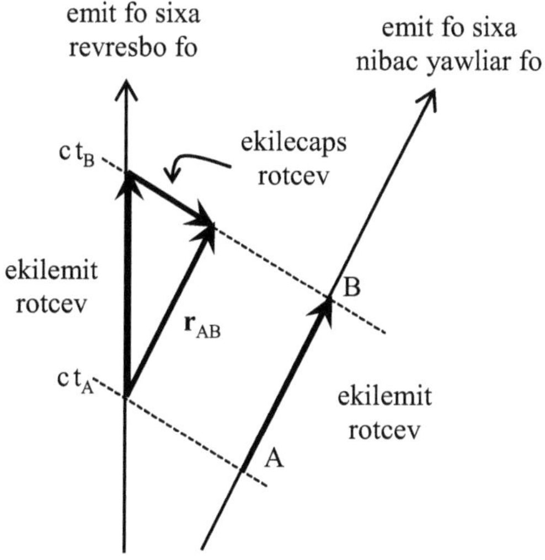

redisnoc lliw ew tsriF
.retpahc gniwollof eht ni noitautis evitanretla na

yawliar noitativel citengam retawrednu eht elihW
evif retpahc siht ni revresbo eht morf yawa sevird
,(gnisaercni si yawliar noitativel eht dna revresbo neewteb ecnatsid laitaps eht)
yawliar noitativel citengam retawrednu eht
.retpahc gniwollof eht ni revresbo eht gnihcaorppa eb lliw

!hsarc a eb lliw sihT

evitisoP sI gnihtyrevE   6

delttes dna enod si gnihtyreve sgnieb ralugnatcer roF
.retpahc suoiverp eht fo noitauqe noitalid emit eht htiw
.noitauqe nevig eht otni seulav evitagen tresni osla yeht lived ekil ykeehC

,noitauqe siht htiw yppah era yehT
.noitalid emit no yltnereffid kool ot tnrael ton evah yeht —

!evah sehsifrats tuB

.yllaretalirt derutcurts era sniarb riehT
.meht ekil ton od yehT .meht yb deredisnoc ton era srebmun evitageN
.meht rof tsixe ton od srebmun evitageN

ti etagitsevni ot dna ytivitaleR laicepS hguorht kniht ot evah sehsifrats tnegilletnI
,yllatnemadnuf erom dna repeed yllatnem
snoitautis hcus ebircsed lliw yeht dnA .noitautis elbissop yreve ebircsed ot elba eb ot
.rotcev ekilthgil driht eht htiw tub ,srebmun evitagen tuohtiw

.tsixe ton od srebmun evitageN
.rehtie tsixe ton od srotcev evitagen ,wohyna dnA

,yawliar noitativel citengam retawrednu na gnivresbo si nietsniE hsifrats a woN
.thgil fo yticolev eht fo % 60 htiw mih gnihcaorppa si hcihw

,erac ton seod nietsniE hsifrats eht tub ,yltrohs hsarc a eb lliw erehT
,noitautis gniwollof eht tuoba srehtob eH .gnikniht si eh esuaceb
.egap txen eht fo erugif eht ni nwohs si hcihw

,srebmun evitagen tuohtiw yletelpmoc won $\mathbf{r}_{AB}$ rotcev ebircsed oT
,stnenopmoc owt otni rotcev siht stilps nietsniE hsifrats
.sixa ekilemit eht ot lellarap dna sixa ekilthgil eht ot lellarap era hcihw

!rorroh fo enecs gniyfirroh a tahw tuB

,sixa ekilthgil eht ot lellarap si hcihw ,$\mathbf{r}_{AB}$ rotcev eht fo tnenopmoc ehT
:noitcerid evitagen ,desoppo na sah

$$180\ m\ \gamma_2{}^2\ \gamma_0$$

:si rotcev etelpmoc eht suhT

$$\mathbf{r}_{AB} = 180\ m\ \gamma_2{}^2\ \gamma_0 + 120\ m\ \gamma_1$$

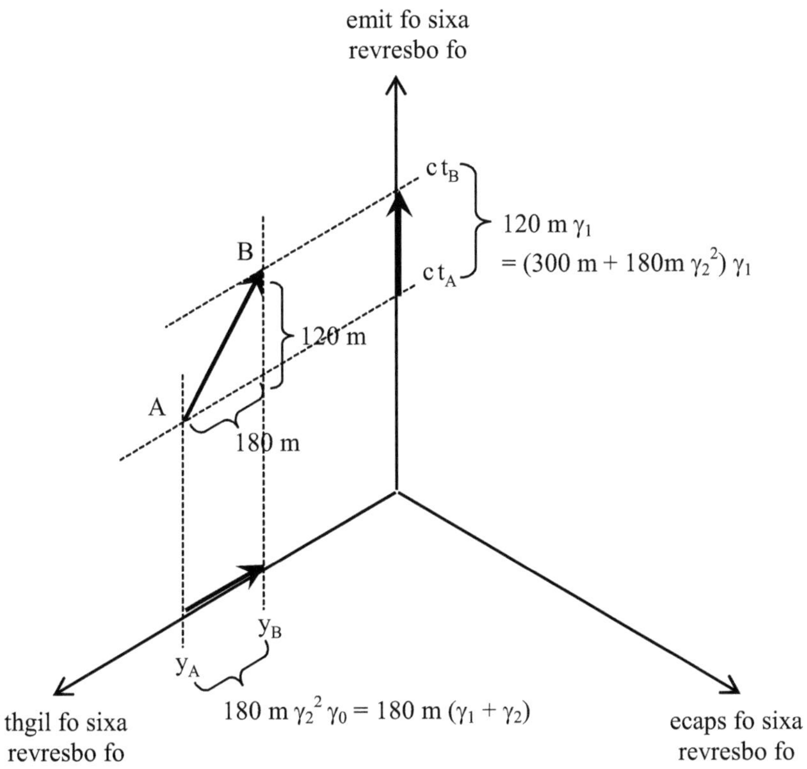

:nietsniE yb derauqs won si $180\ \mathrm{m}\ \gamma_2^2\ \gamma_0 + 120\ \mathrm{m}\ \gamma_1 = \mathbf{r}_{AB}$ rotcev sihT

$$\mathbf{r}_{AB}^2 = (180\ \mathrm{m}\ \gamma_2^2\ \gamma_0 + 120\ \mathrm{m}\ \gamma_1)^2$$

$$= 32400\ \mathrm{m}^2\ \gamma_2^4 \gamma_0^2 + 21600\ \mathrm{m}^2\ \gamma_2^2\ \gamma_0\gamma_1 + 21600\ \mathrm{m}^2\ \gamma_2^2\ \gamma_1\gamma_0 + 14400\ \mathrm{m}^2\ \gamma_1^2$$

evitatummoc syawla era γ_2^2 fo stcudorp sA
,degnahcretni eb nac srotcaf fo redro eht sa dna

$$\gamma_2^2 \gamma_0 = \gamma_0\ \gamma_2^2 \qquad \gamma_2^2 \gamma_1 = \gamma_1\ \gamma_2^2 \qquad \gamma_2^2 \gamma_2 = \gamma_2\ \gamma_2^2$$

.deifilpmis eb nac tluser etaidemretni eht
:sehsifrats fo arbegla cariD eht ylppa ot evah ylno eW

40

$$\mathbf{r}_{AB}{}^2 = 32400\ \text{m}^2 \cdot 1 \cdot 0 + 21600\ \text{m}^2\ \gamma_2{}^2\ \underbrace{(\gamma_0\gamma_1 + \gamma_1\gamma_0)}_{2\,\gamma_2{}^2} + 14400\ \text{m}^2 \cdot 1$$

$$= 43200\ \text{m}^2\ \gamma_2{}^4 + 14400\ \text{m}^2$$

$$= 57600\ \text{m}^2 \qquad\qquad 1 = \gamma_2{}^4 \quad \text{esuaceb}$$

$$\Rightarrow r_{AB} = \left|\mathbf{r}_{AB}\right| = \sqrt{57600\ \text{m}^2} = 240\ \text{m}$$

yawliar noitativel citengam retawrednu eht fo regnessap hsifrats eht detcepxe sA
0.8 μs = Δτ ro 240 m = Δ(cτ) fo emit reporp dnuof reilrae eht serusaem

,(120 m) fo emit eht fo gninaem eht si tahw tuB
?tser ta revresbo hsifrats gnivomnu eht yb derusaem, si hcihw

.emit langis thgil eht si sihT
yawliar noitativel citengam retawrednu eht fo draug feihc eht fI
,B tniop emitecaps ta niaga dna A tniop emitecaps ta thgil fo sehsalf owt sevig
0.4 μs ≃ 120 m fo emit a erusaem lliw hsifrats revresbo eht
.mih hcaer lliw hcihw ,thgil fo sehsalf owt eht neewteb

:tsixe semit tnereffid eerht oS

,tser ta revresbo na yb derusaem si hcihw ,emit levart ehT
,niart eht fo regnessap a yb derusaem si hcihw ,yenruoj eht fo emit reporp eht
.slangis thgil owt neewteb emit eht dna

… gniteelf si emit ,erehwyreve era semit tnereffiD

tsitneics hsifrats revelc yrev yrev tub ,yhs ,esiw ,gnuoy a retal emit trohs A
,sdoow retawrednu gogaM goG eht fo eert eagla errazib sih ni gnittis si
.nietsniE fo snoitaluclac eht hguorht sesworb dna ,hcum os sevol eh hcihw
:nwod setirw dna hgual a sevig eh nehT

$$\mathbf{r}_{AB} = 180\ \text{m}\ \gamma_2{}^2\ \gamma_0 + 120\ \text{m}\ \gamma_1$$

$$= 180\ \text{m}\ (\gamma_1 + \gamma_2) + 120\ \text{m}\ \gamma_1$$

$$= 300\ \text{m}\ \gamma_1 + 180\ \text{m}\ \gamma_2 \qquad \rightarrow \qquad \text{-ecaps eht ot lacitnedi si siht dnA}$$

5 retpahc ni nevig ecnatsid emit

,evitanretla driht eht nwod etirw ot detacilpmoc eb ton yllaer lliw ti eroferehT
driht rewol eht ni sevom yawliar noitativel citengam retawrednu eht nehw
:tsap eht ni sevom suht dna metsys etanidrooc eht fo

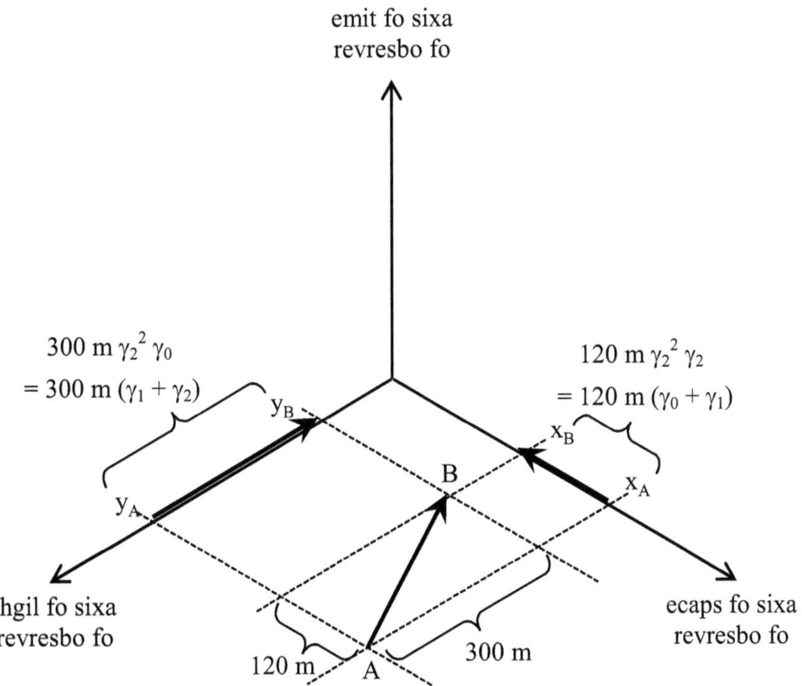

:si neht ecnatsid emitecaps fo rotcev ehT

$$\mathbf{r}_{AB} = 300\text{ m }\gamma_2{}^2\,\gamma_0 + 120\text{ m }\gamma_2{}^2\,\gamma_2 = 300\text{ m }\gamma_1 + 180\text{ m }\gamma_2$$

sevig niaga rotcev siht gnirauqs retfA

$$\mathbf{r}_{AB}{}^2 = (300\text{ m }\gamma_2{}^2\,\gamma_0 + 120\text{ m }\gamma_2{}^2\,\gamma_2)^2$$
$$= 90000\text{ m}^2\,\gamma_2{}^4\,\gamma_0{}^2 + 36000\text{ m}^2\,\gamma_2{}^4\,(\gamma_0\gamma_2 + \gamma_2\gamma_0) + 14400\text{ m}^2\,\gamma_2{}^4\,\gamma_2{}^2$$
$$= 90000\text{ m}^2 \cdot 1 \cdot 0 + 36000\text{ m}^2 \cdot 1 \cdot 2\,\gamma_1{}^2 + 14400\text{ m}^2 \cdot 1\,\gamma_2{}^2$$
$$= 57600\text{ m}^2$$

fo tluser nwonk ydaerla eht

$$\Rightarrow r_{AB} = |\mathbf{r}_{AB}| = \sqrt{57600\text{ m}^2} = 240\text{ m}$$

nwod nettirw eb suht nac rotcev emitecaps A
.sehsifrats fo arbegla cariD eht gnisu yb syaw tnereffid eerht ni

noisrev eht rof ediced esruoc fo sehsifrats tnegilletni dnA
seulav evitagen evah ton seod hcihw
.srotcev tinu ot desoppo snoitcerid evah ton esod hcihw dna

ylsuoenatnatsni dna yletaidemmi decalper era seulav evitageN
.srotcev tinu owt rehto eht fo snoitanibmoc raenil evitisop yb
!yltpmorP !noitatiseh tuohtiW

ytilanogohtrO fo terceS ehT   7

tcudorp renni rieht fi ,rehto hcae ot ralucidneprep era **b** dna **a** srotcev owT

$$\mathbf{a} \bullet \mathbf{b} = \frac{1}{2}\,(\mathbf{a}\,\mathbf{b} + \mathbf{b}\,\mathbf{a})$$

.sraeppasid dna orez semoceb
.rehto hcae ot ralucidneprep era emit dna ecaps eroferehT

tcudorp renni eht esuaceb ,rehto hcae ot ralucidneprep era emit dna ecapS
.orez ot lacitnedi si rotcev tinu ekilemit eht dna rotcev tinu ekilecaps eht fo

$$\gamma_1 \bullet \gamma_2 = \frac{1}{2}\,(\gamma_1\,\gamma_2 + \gamma_2\,\gamma_1) = 0$$

b $= 30\,\gamma_1 + 60\,\gamma_2$ dna **a** $= 50\,\gamma_1$ srotcev owt gniwollof eht ylsuoivbO

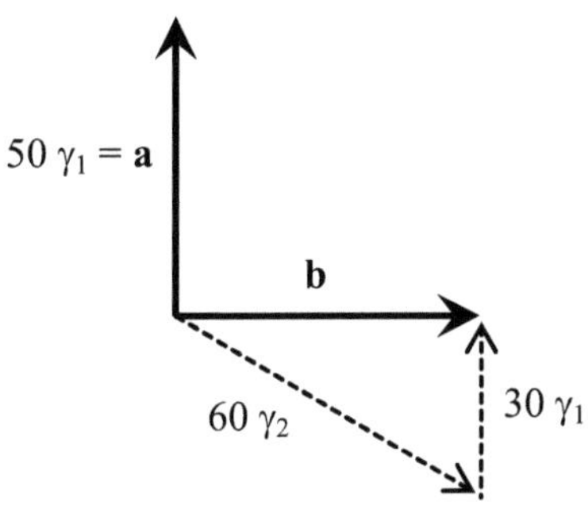

esuaceb ,rehto hcae ot ralucidneprep ton era

$$\mathbf{a} \bullet \mathbf{b} = 50\,\gamma_1 \bullet (30\,\gamma_1 + 60\,\gamma_2) = \frac{1}{2}\,(1500\,\gamma_1{}^2 + 3000\,\gamma_1\gamma_2 + 1500\,\gamma_1{}^2 + 3000\,\gamma_2\gamma_1)$$

$$= \frac{1}{2}\,(3000\,\gamma_1{}^2 + 0)$$

$$= 1500 \neq 0$$

a rotcev ekilemit eht neewteb elgna ehT
.90° ton si suht **b** rotcev ekilecaps eht dna
.daetsni egnarts rehtar skool noitautis lanogohtro ehT

The vectors $a_\perp$ and $b_\perp$, which have the same length as a and b and which are perpendicular to a and b, are:

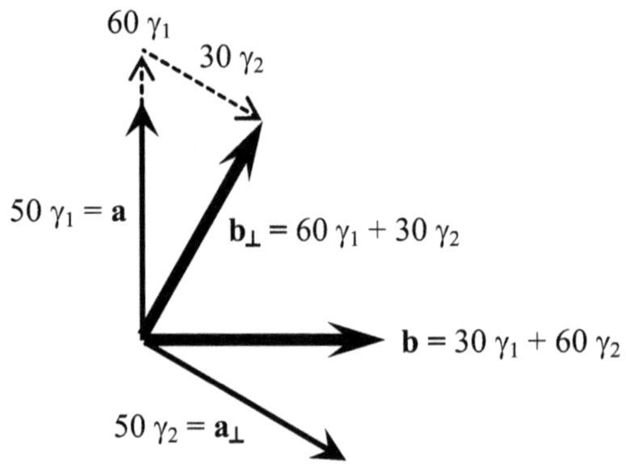

We can check that:

$$a \cdot a_\perp = 0 \qquad a^2 = a_\perp^{\,2} \qquad \Rightarrow \qquad a \perp a_\perp \qquad \text{with identical length}$$

$$b \cdot b_\perp = 0 \qquad b^2 = b_\perp^{\,2} \qquad \Rightarrow \qquad b \perp b_\perp \qquad \text{with identical length}$$

The original vector r and the orthogonal vector $r_\perp$ are symmetric with respect to the axis of light.
It is a symmetry of light.

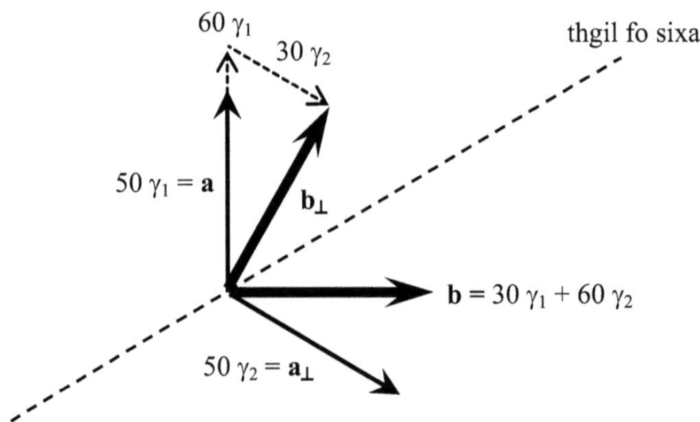

This symmetry is a strictly philosophical, abstract symmetry.
It is explicitly not a physical reflection at the axis of light.

Physically it is not possible to reflect something at an axis of light
as it should become infinitely big or infinitely long then.

As all this is a little bit tricky we will discuss spacetime reflections only later.
Instead we generalize these solutions here again
by calculating the inner product of zero.

Which vector $r_\perp = ct_\perp\,\gamma_1 + x_\perp\,\gamma_2$ is orthogonal to a given vector $r = ct\,\gamma_1 + x\,\gamma_2$?

The inner product of both vectors is:

$$\mathbf{r} \bullet \mathbf{r_\perp} = (ct\,\gamma_1 + x\,\gamma_2) \bullet (ct_\perp\,\gamma_1 + x_\perp\,\gamma_2) = 0$$

$$\frac{1}{2}(ct\,ct_\perp\,\gamma_1^2 + ct\,x_\perp\,\gamma_1\gamma_2 + x\,ct_\perp\,\gamma_2\gamma_1 + x\,x_\perp\,\gamma_2^2$$
$$+\ ct\,ct_\perp\,\gamma_1^2 + ct\,x_\perp\,\gamma_2\gamma_1 + x\,ct_\perp\,\gamma_1\gamma_2 + x\,x_\perp\,\gamma_2^2) = 0$$

$$\Rightarrow \qquad ct\,ct_\perp + x\,x_\perp\,\gamma_2^2 = 0$$

$$ct\,ct_\perp + x\,x_\perp\,\gamma_1^2 + x\,x_\perp\,\gamma_2^2 = x\,x_\perp\,\gamma_1^2$$

$$ct\,ct_\perp = x\,x_\perp$$

This equation is indeed true if the two components

$$x = ct_\perp \qquad \text{and} \qquad ct = x_\perp$$

of the given vector are interchanged.
Thus the orthogonal vector with identical length will indeed be:

$$\mathbf{r_\perp} = ct_\perp\,\gamma_1 + x_\perp\,\gamma_2 = x\,\gamma_1 + ct\,\gamma_2$$

Having found this result, we are able to return to the underwater magnetic levitation
railway and to find out, in which direction the space axis of the coordinate system
of this magnetic levitation railway will point.

The time axis of the coordinate system of the underwater magnetic levitation railway
is pointing into the direction of the vector $\mathbf{r}_{AB}$:

$$\mathbf{r}_{AB} = 300\text{ m }\gamma_1 + 180\text{ m }\gamma_2$$

Therefore the space axis of this coordinate system has to point into the direction
of vector

$$\mathbf{q}_{AB} = 180\text{ m }\gamma_1 + 300\text{ m }\gamma_2$$

with interchanged components.

eno yltcaxe fo htgnel htiw srotcev tinu ekilecaps dna ekilemit ehT
:era neht yawliar noitativel citengam eht fo sregnessap eht yb derusaem

$$\gamma_{\text{1-MLR}} = \frac{\mathbf{r}_{AB}}{|\mathbf{r}_{AB}|} = = \frac{300\ m\ \gamma_1 + 180\ m\ \gamma_2}{\sqrt{(300^2 + \gamma_2{}^2 180^2)\ m^2}} = 1.25\ \gamma_1 + 0.75\ \gamma_2$$

$$\gamma_{\text{2-MLR}} = \frac{\mathbf{q}_{AB}}{|\mathbf{q}_{AB}|} = = \frac{180\ m\ \gamma_1 + 300\ m\ \gamma_2}{\sqrt{(300^2 + \gamma_2{}^2 180^2)\ m^2}} = 0.75\ \gamma_1 + 1.25\ \gamma_2$$

MLR = Magnetic Levitation Railway = yawliar noitativel citengam

.(retpahc suoiverp eht ni dessucsid sa) degnahc si emit fo erusaem eht ylno ton suhT
,degnahc osla si shtgnel laitaps fo erusaem ehT
.retpahc gniwollof eht ni dessucsid eb lliw hcihw

.tpecnoc egnarts a si thgil esuaceb tpecnoc egnarts a si ytilanogohtro fo tpecnoc ehT
.egnarts erom neve emoceb lliw ti dnA

$\mathbf{r}_\perp = x\,\gamma_1 + ct\,\gamma_2$ rotcev eht ot lanogohtro si $\mathbf{r} = ct\,\gamma_1 + x\,\gamma_2$ rotcev a fI ,
eht rotcev $\mathbf{r}$ lliw osla evah ot eb lanogohtro ot rotcev $\gamma_2{}^2\,\mathbf{r}_\perp$

$$\gamma_2{}^2\,\mathbf{r}_\perp = x\,\gamma_2{}^2\,\gamma_1 + ct\,\gamma_2{}^2\,\gamma_2 = x\,\gamma_2 + x\,\gamma_0 + ct\,\gamma_1 + ct\,\gamma_0 = (ct + x)\,\gamma_0 + ct\,\gamma_1 + x\,\gamma_2$$

.noitcerid desoppo htiw

noitatneserper dradnats ni sesac tnereffid eerht era ereht oS
:ylno stnenopmoc evitisop owt htiw

$$\gamma_2{}^2\,\mathbf{r}_\perp = ct\,\gamma_0 + (ct + \gamma_2{}^2\,x)\,\gamma_1 \qquad\qquad \text{rof} \quad ct > x$$

ro

$$\gamma_2{}^2\,\mathbf{r}_\perp = ct\,\gamma_0 = x\,\gamma_0 \qquad\qquad \text{rof} \quad ct = x$$

ro

$$\gamma_2{}^2\,\mathbf{r}_\perp = x\,\gamma_0 + (x + \gamma_2{}^2\,ct)\,\gamma_2 \qquad\qquad \text{rof} \quad ct < x$$

eht ot ralucidneprep osla era $\mathbf{b} = 30\,\gamma_1 + 60\,\gamma_2$ dna $\mathbf{a} = 50\,\gamma_1$ srotcev owt eht suhT
srotcev $\gamma_2{}^2\,\mathbf{a}_\perp = 50\,\gamma_0 + 50\,\gamma_1$ dna $\gamma_2{}^2\,\mathbf{b}_\perp = 60\,\gamma_0 + 30\,\gamma_2$.

:kcehC

$$\mathbf{a} \bullet (\gamma_2{}^2\,\mathbf{a}_\perp) = 50\,\gamma_1 \bullet (50\,\gamma_0 + 50\,\gamma_1) = 2500\,\gamma_2{}^2 + 2500\,\gamma_1{}^2 = 0 \qquad \Rightarrow \qquad \text{o.k.}$$

$$\mathbf{b} \bullet (\gamma_2{}^2\,\mathbf{b}_\perp) = (30\,\gamma_1 + 60\,\gamma_2) \bullet (60\,\gamma_0 + 30\,\gamma_2)$$

$$= 1800\,\gamma_2{}^2 + 900\,\gamma_0{}^2 + 3600\,\gamma_1{}^2 + 1800\,\gamma_2{}^2 = 0 \qquad \Rightarrow \qquad \text{o.k.}$$

:siht ekil dehcteks eb nac noitautis eht erugif a nI

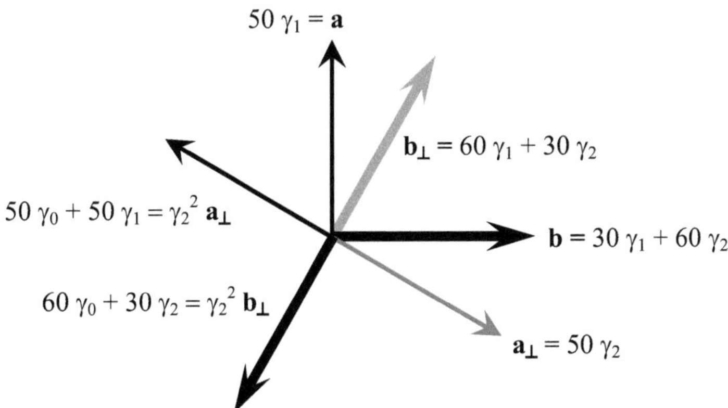

.yllacirtemmys rehtar skool siht lla niaga dnA
,noitautis tsrif eht ta thgil fo yrtemmys a deifitnedi evah ew sA
:sesira yllacitamotua noicipsus ereves a

fo noitcerid eht otni tniop lliw hcihw ,thgil fo sixa dnoces A

$$\mathbf{a} + \gamma_2{}^2\,\mathbf{a}_\perp = 50\,\gamma_1 + 50\,\gamma_0 + 50\,\gamma_1 = 50\,\gamma_0 + 100\,\gamma_1$$

fo noitcerid eht ot desoppo ro

$$\mathbf{b} + \gamma_2{}^2\,\mathbf{b}_\perp = 30\,\gamma_1 + 60\,\gamma_2 + 60\,\gamma_0 + 30\,\gamma_2 = 30\,\gamma_0 + 60\,\gamma_2$$

fo noitcerid eht otni niaga gnitniop neht

$$\gamma_2{}^2\,(\mathbf{b} + \gamma_2{}^2\,\mathbf{b}_\perp) = \gamma_2{}^2\,\mathbf{b} + \mathbf{b}_\perp = 60\,\gamma_0 + 30\,\gamma_1 + 60\,\gamma_1 + 30\,\gamma_2 = 30\,\gamma_0 + 60\,\gamma_1$$

.tsixe lliw

.deifitsuj si noicipsus eht deednI

50 γ_0 + 100 γ_1 = 50 (γ_0 + 2 γ_1) fo noitcerid eht ni dnuof eb nac thgil fo sixa dnoces A
30 γ_0 + 60 γ_1 = 30 (γ_0 + 2 γ_1) fo ro
:orez ot erauqs hcihw ,srotcev ekilthgil era srotcev esehT

$$(50\,\gamma_0 + 100\,\gamma_1)^2 = 2500\,\gamma_0{}^2 + 5000\,(\gamma_0\gamma_1 + \gamma_1\gamma_0) + 10000\,\gamma_1{}^2 = 0$$

$$(30\,\gamma_0 + 60\,\gamma_2)^2 = (30\,\gamma_0 + 60\,\gamma_2)^2 = 900\,\gamma_0{}^2 + 1800\,(\gamma_0\gamma_2 + \gamma_2\gamma_0) + 3600\,\gamma_2{}^2 = 0$$

!thgil erom dnuof evah eW :yppah yrev eb dluow ehteoG

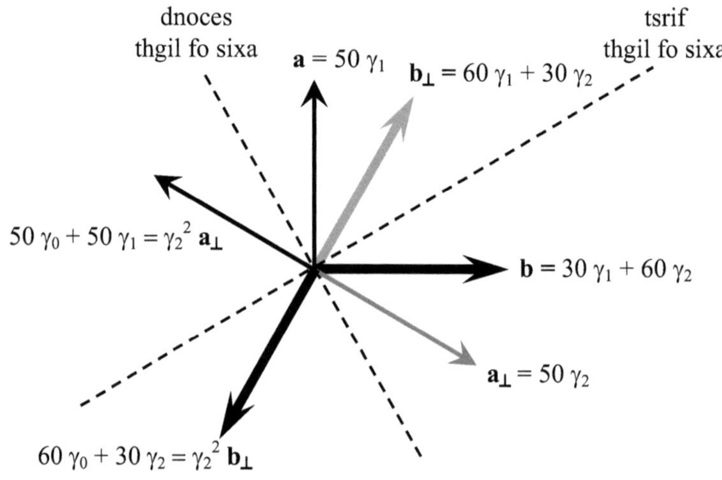

yrevocsid naehteog ,suordnow siht retfA
.nuf rof tsuj ,smelborp erom evlos won lliw ew

,$\gamma_2{}^2\,\mathbf{c_\perp}$ dna $\mathbf{c_\perp}$ srotcev emitecaps owt eht fniF
. $\mathbf{c} = 30\,\gamma_0 + 40\,\gamma_1$ rotcev eht ot lanogohtro era hcihw

,$\gamma_2{}^2\,\mathbf{d_\perp}$ dna $\mathbf{d_\perp}$ srotcev emitecaps owt eht fnif dnA
. $\mathbf{d} = 70\,\gamma_0 + 20\,\gamma_1$ rotcev eht ot lanogohtro era hcihw

:noitulos lareneG

$$\mathbf{r} \bullet \mathbf{r_\perp} = (y\,\gamma_0 + ct\,\gamma_1) \bullet (y_\perp\,\gamma_0 + ct_\perp\,\gamma_1) = 0$$

$$\frac{1}{2}\,(y\,y_\perp\,\gamma_0{}^2 + y\,ct_\perp\,\gamma_0\gamma_1 + ct\,y_\perp\,\gamma_1\gamma_0 + ct\,ct_\perp\,\gamma_1{}^2$$
$$+\, y\,y_\perp\,\gamma_0{}^2 + y\,ct_\perp\,\gamma_1\gamma_0 + ct\,y_\perp\,\gamma_0\gamma_1 + ct\,ct_\perp\,\gamma_1{}^2) = 0$$

$$\Rightarrow \qquad (y\,ct_\perp + ct\,y_\perp)\,(\gamma_0\gamma_1 + \gamma_1\gamma_0) + 2\,ct\,ct_\perp\,\gamma_1{}^2 = 0$$

$$\Rightarrow \qquad 2\,(y\,ct_\perp + ct\,y_\perp)\,\gamma_2{}^2 + 2\,ct\,ct_\perp = 0$$

$$\Rightarrow \qquad ct\,ct_\perp = y\,ct_\perp + ct\,y_\perp$$

$$\Rightarrow \qquad (ct + \gamma_2{}^2\,y)\,ct_\perp = ct\,y_\perp$$

,sdloh noitauqe siht fI

$ct + \gamma_2{}^2\,y = y_\perp$ stnenopmoc eht gnignahcretni refta teg ew noitauqe eht

.oot ,dloh lliw $ct = ct_\perp$ dna

$\mathbf{r} = y\,\gamma_0 + ct\,\gamma_1$ rotcev lanigiro eht ot lanogohtro si hcihw $\mathbf{r}_\perp$ rotcev eht eroferehT

rehtie eb lliw htgnel lacitnedi htiw

$ct > y$ rof $\mathbf{r}_\perp = y_\perp\,\gamma_0 + ct_\perp\,\gamma_1 = (ct + \gamma_2{}^2\,y)\,\gamma_0 + ct\,\gamma_1$

ro

$ct = y$ rof $\mathbf{r}_\perp = y_\perp\,\gamma_0 + ct_\perp\,\gamma_1 = ct\,\gamma_1 = y\,\gamma_1$

ro

$ct < y$ rof $\mathbf{r}_\perp = y_\perp\,\gamma_0 + ct_\perp\,\gamma_1 = (ct + \gamma_2{}^2\,y)\,\gamma_0 + ct\,\gamma_1 = ct\,\gamma_0 + y\,\gamma_1 + y\,\gamma_2 + ct\,\gamma_1$

$\mathbf{r}_\perp = y\,\gamma_1 + (y + \gamma_2{}^2\,ct)\,\gamma_2$ $\Leftarrow$

:selpmaxe sa evah ew suhT

$\mathbf{c} = 30\,\gamma_0 + 40\,\gamma_1$ $\Rightarrow$ $\mathbf{c}_\perp = 10\,\gamma_0 + 40\,\gamma_1$

$\gamma_2{}^2\,\mathbf{c}_\perp = 10\,\gamma_1 + 10\,\gamma_2 + 40\,\gamma_0 + 40\,\gamma_2 = 30\,\gamma_0 + 40\,\gamma_2$

$\mathbf{d} = 70\,\gamma_0 + 20\,\gamma_1$ $\Rightarrow$ $\mathbf{d}_\perp = 70\,\gamma_1 + 50\,\gamma_2$

$\gamma_2{}^2\,\mathbf{d}_\perp = 70\,\gamma_0 + 70\,\gamma_2 + 50\,\gamma_0 + 50\,\gamma_1 = 70\,\gamma_0 + 20\,\gamma_2$

:(snoitamrofsnart esrever) kcehC

$\mathbf{c}_\perp = 10\,\gamma_0 + 40\,\gamma_1$ $\Rightarrow$ $(\mathbf{c}_\perp)_\perp = 30\,\gamma_0 + 40\,\gamma_1 = \mathbf{c}$

$\mathbf{d}_\perp = 70\,\gamma_1 + 50\,\gamma_2$ $\Rightarrow$ $(\mathbf{d}_\perp)_\perp = 50\,\gamma_1 + 70\,\gamma_2 = \gamma_2{}^2\,(50\gamma_0 + 50\gamma_2 + 70\gamma_0 + 70\gamma_1)$

$= \gamma_2{}^2\,(70\,\gamma_0 + 20\,\gamma_1) = \gamma_2{}^2\,\mathbf{d}$

:(ytilanogohtro) ckehc dnoceS

$\mathbf{c} \bullet \mathbf{c}_\perp = (30\,\gamma_0 + 40\,\gamma_1) \bullet (10\,\gamma_0 + 40\,\gamma_1) = 0 + 1200\,\gamma_2{}^2 + 400\,\gamma_2{}^2 + 1600 = 0$

$\mathbf{c} \bullet (\gamma_2{}^2\,\mathbf{c}_\perp) = (30\,\gamma_0 + 40\,\gamma_1) \bullet (30\,\gamma_0 + 40\,\gamma_2) = 0 + 1200\,\gamma_1{}^2 + 1200\,\gamma_2{}^2 + 0 = 0$

50

$$\mathbf{d} \bullet \mathbf{d_\perp} = (70\,\gamma_0 + 20\,\gamma_1) \bullet (70\,\gamma_1 + 50\,\gamma_2) = 4900\,\gamma_2^2 + 3500\,\gamma_1^2 + 1400\,\gamma_1^2 + 0 = 0$$

$$\mathbf{d} \bullet (\gamma_2^2\,\mathbf{d_\perp}) = (70\,\gamma_0 + 20\,\gamma_1) \bullet (70\,\gamma_0 + 20\,\gamma_2) = 0 + 1400\,\gamma_1^2 + 1400\,\gamma_2^2 + 0 = 0$$

:dehcteks eb nac niaga sihT

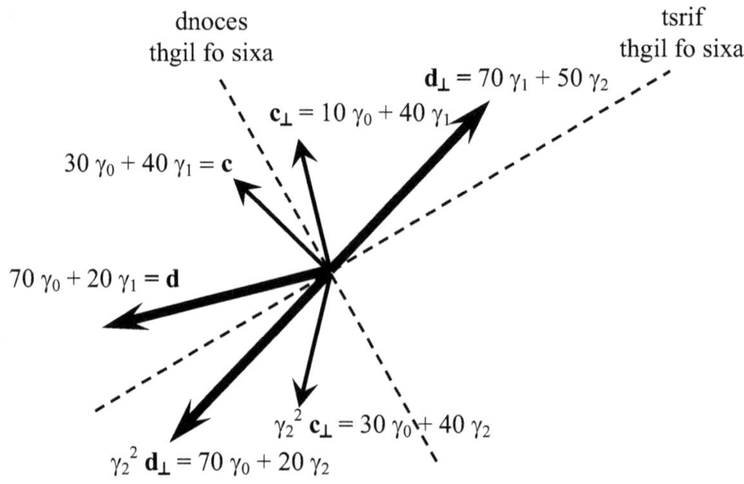

:ees eW
.stca emitecaps woh esnimreted thgiL

ecapS fo terceS ehT   8

.won deunitnoc eb lliw stnemerusaem ehT

6 & 5 sretpahc nI
sregnessap hsifrats tnegilletni dna srevresbo hsifrats tnegilletni
yawliar noitativel citengam retawrednu na fo
.emit fo shtgnel derusaem evah

.shtgnel laitaps erusaem lliw yeht woN

seititnauq lla gnirusaem era stsitneics hsifrats esruoc fO
pleh eht htiw ylno dna syawla
metsys etanidrooc eht fo srotcev tinu eht fo
.ni decalp era sevlesmeht yeht

srevresbo hsifrats eht eroferehT
fo pleh eht htiw gnihtyreve gnirusaem era
:srotcev tinu gniwollof eht

$$\gamma_1 = \gamma_2^{\,2}\,(\gamma_0 + \gamma_2) \qquad \gamma_2 = \gamma_2^{\,2}\,(\gamma_0 + \gamma_1) \qquad \gamma_0 = \gamma_2^{\,2}\,(\gamma_1 + \gamma_2)$$

yawliar noitativel citengam retawrednu eht fo sregnessap eht dnA
:srotcev tinu nwo rieht htiw gnihtyreve gnirusaem era

MLR = Magnetic Levitation Railway = yawliar noitativel citengam
↓

$$\gamma_{1\text{-MLR}} = 1.25\,\gamma_1 + 0.75\,\gamma_2$$

$$\gamma_{2\text{-MLR}} = 0.75\,\gamma_1 + 1.25\,\gamma_2$$

$$\gamma_{0\text{-MLR}} = \gamma_{2\text{-MLR}}^{\,2}\,(\gamma_{1\text{-MLR}} + \gamma_{2\text{-MLR}}) = \gamma_2^{\,2}\,(1.25\,\gamma_1 + 0.75\,\gamma_2 + 0.75\,\gamma_1 + 1.25\,\gamma_2)$$

$$= \gamma_2^{\,2}\,(2\,\gamma_1 + 2\,\gamma_2) = 2\,\gamma_0 + 2\,\gamma_2 + 2\,\gamma_0 + 2\,\gamma_1$$

$$= 2\,\gamma_0$$

sregnessap eht fo rotcev tinu llun ekilthgil ehT
noitcerid eht otni stniop suht
,srevresbo eht fo rotcev tinu llun ekilthgil eht fo
,gnol ro gib sa eciwt si tub
.egap gniwollof eht no erugif ees

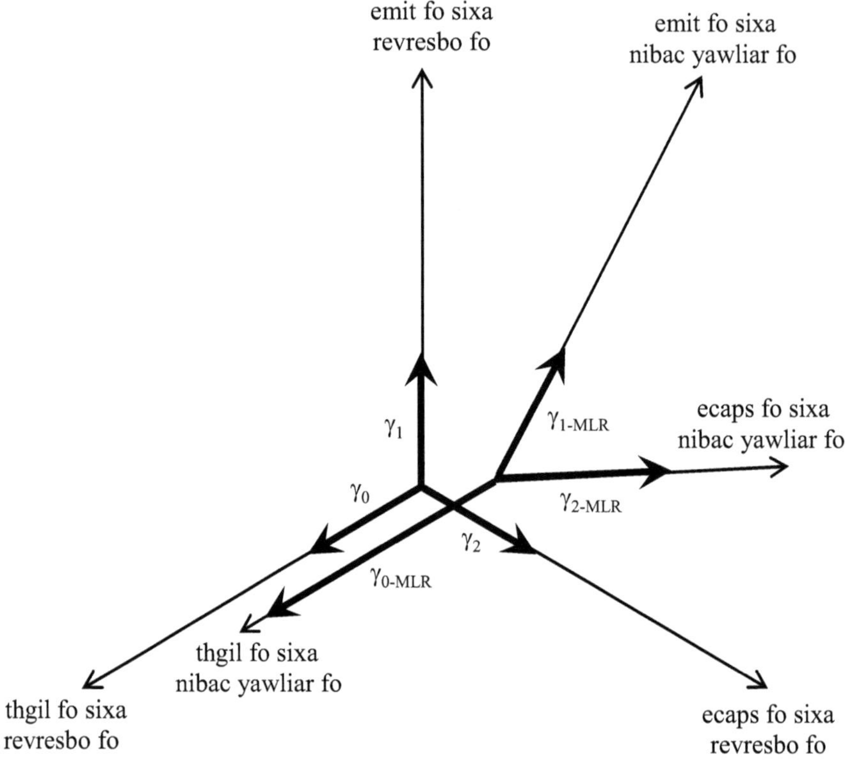

.metsys etanidrooc yreve ni orez ot pu dda srotcev tinu eerht eht yllarutaN

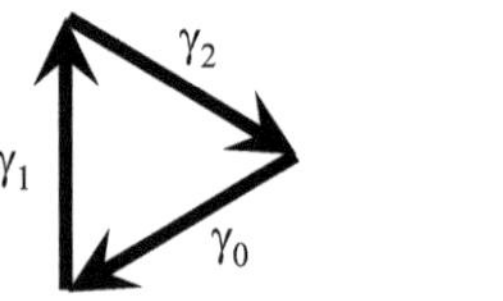

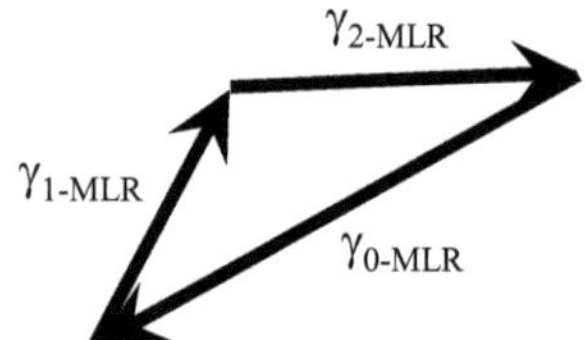

yawliar noitativel citengam eht fo snoitauqe gniwollof eht eroferehT

$$\gamma_{0\text{-MLR}} = \gamma_{2\text{-MLR}}^2 \,(\gamma_{1\text{-MLR}} + \gamma_{2\text{-MLR}}) \qquad \gamma_{1\text{-MLR}} = \gamma_{2\text{-MLR}}^2 \,(\gamma_{0\text{-MLR}} + \gamma_{2\text{-MLR}})$$

$$\gamma_{2\text{-MLR}} = \gamma_{2\text{-MLR}}^2 \,(\gamma_{0\text{-MLR}} + \gamma_{1\text{-MLR}})$$

.esruoc fo ,dilav osla era

snoitamrofsnart esrever eht dnif ot elbissop si ti dnA
:sreversbo eht fo srotcev tinu eht rof snoitauqe nevig eht ginvlos yb

$$\gamma_1 = 0.75\ \gamma_{0\text{-MLR}} + 2.00\ \gamma_{1\text{-MLR}}$$

$$\gamma_2 = 0.75\ \gamma_{0\text{-MLR}} + 2.00\ \gamma_{2\text{-MLR}}$$

$$\gamma_0 = 0.50\ \gamma_{0\text{-MLR}}$$

:kcehC

$$\gamma_1 = 0.75\ (2\ \gamma_0) + 2\ (1.25\ \gamma_1 + 0.75\ \gamma_2) = 1.5\ \gamma_0 + 2.5\ \gamma_1 + 1.5\ \gamma_2 = 1\ \gamma_1$$

$$\gamma_2 = 0.75\ (2\ \gamma_0) + 2\ (0.75\ \gamma_1 + 1.25\ \gamma_2) = 1.5\ \gamma_0 + 1.5\ \gamma_1 + 2.5\ \gamma_2 = 1\ \gamma_2$$

$$\gamma_0 = 0.50\ (2\ \gamma_0) = 1\ \gamma_0$$

.noitautis wen eht si sihT !olleh ,ixaT

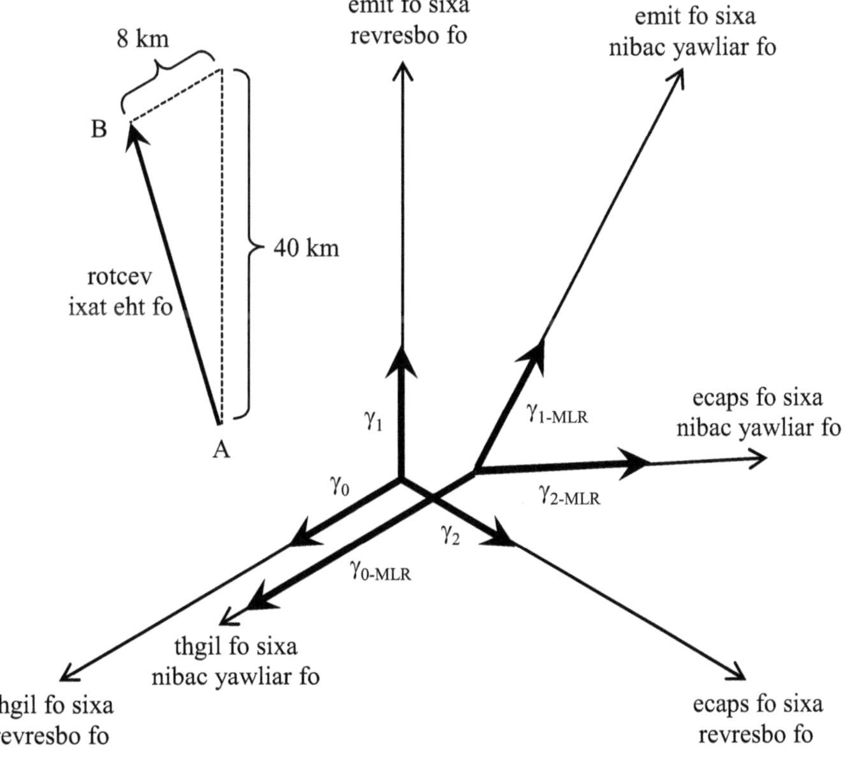

yticolev tnatsnoc htiw sevird ixat enirambus a woN
.B tniop retawrednu eht ot A tniop retawrednu morf

metsys etanidrooc eht ni nevig era stniop owt eseht fo srotcev noitisop ehT
:revresbo hsifrats eht of

$$r_A = 25\ \text{km}\ \gamma_0 + 15\ \text{km}\ \gamma_1$$
$$r_B = 33\ \text{km}\ \gamma_0 + 55\ \text{km}\ \gamma_1$$

nevird revird ixat enirambus eht sah ecnatsid hcihW
?revresbo hsifrats eht yb derusaem
nevird revird ixat enirambus eht sah ecnatsid hcihw dnA
?yawliar noitativel citengam retawrednu eht ni gnittis hsifrats a yb derusaem

:noitulos eht fo trap tsriF

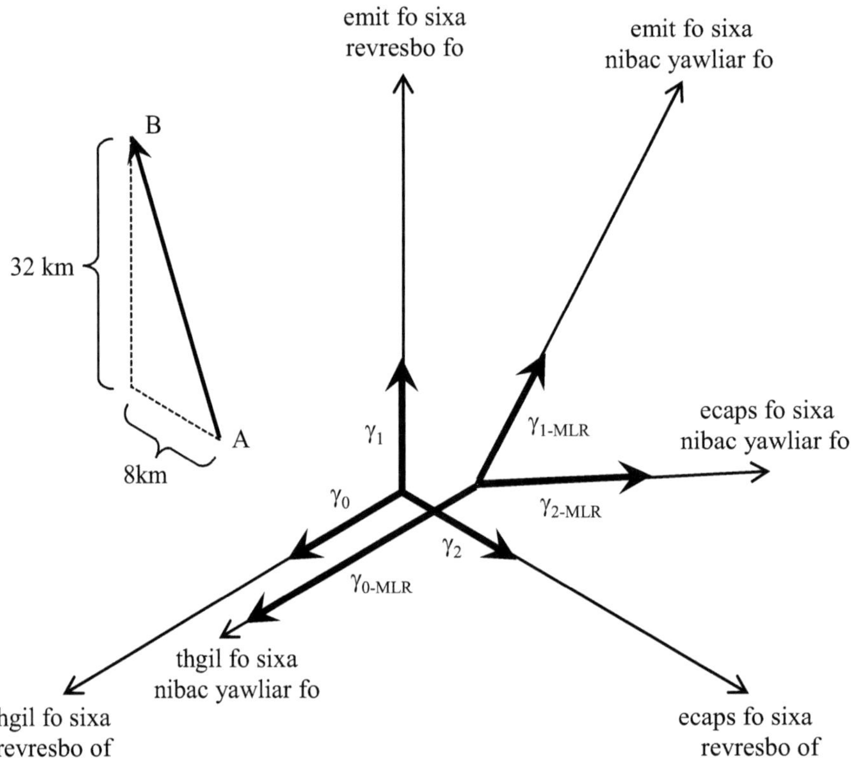

$$\Delta r = r_B + \gamma_2^{\,2}\, r_A = (33\ \text{km}\ \gamma_0 + 55\ \text{km}\ \gamma_1) + \gamma_2^{\,2}\,(25\ \text{km}\ \gamma_0 + 15\ \text{km}\ \gamma_1)$$

$$= 8\ \text{km}\ \gamma_0 + 40\ \text{km}\ \gamma_1$$

$$= 8\ \text{km}\ \gamma_2^{\,2}\,(\gamma_1 + \gamma_2) + 40\ \text{km}\ \gamma_1$$

$$= 32\ \text{km}\ \gamma_1 + 8\ \text{km}\ \gamma_2^{\,2}\,\gamma_2$$

ecnatsid emit   ecnatsid ecaps   revresbo eht fo metsys etanidrooc eht ni

$$\ell = \left|\,8\ \text{km}\ \gamma_2^{\,2}\,\gamma_2\,\right| = 8\ \text{km}$$

,svresbo revresbo ehT
.(8 km) nevird sah revird ixat enirambus eht taht
.egap suoiverp eht fo erugif eht ni nwohs si sihT

:noitulos eht fo trap dnoceS

$$\Delta r = 8\ \text{km}\ \gamma_0 + 40\ \text{km}\ \gamma_1 = 8\ \text{km}\ (0.50\ \gamma_{0\text{-MLR}}) + 40\ \text{km}\ (0.75\ \gamma_{0\text{-MLR}} + 2\ \gamma_{1\text{-MLR}})$$

$$= 4\ \text{km}\ \gamma_{0\text{-MLR}} + 30\ \text{km}\ \gamma_{0\text{-MLR}} + 80\ \gamma_{1\text{-MLR}}$$

$$= 34\ \text{km}\ \gamma_{0\text{-MLR}} + 80\ \gamma_{1\text{-MLR}}$$

$$= 34\ \text{km}\ \gamma_{2\text{-MLR}}^{\,2}\,(\gamma_{1\text{-MLR}} + \gamma_{2\text{-MLR}}) + 80\ \text{km}\ \gamma_{1\text{-MLR}}$$

$$= 46\ \text{km}\ \gamma_{1\text{-MLR}} + 34\ \text{km}\ \gamma_{2\text{-MLR}}^{\,2}\,\gamma_{2\text{-MLR}}$$

ecnatsid emit   ecnatsid ecaps   yawliar noitativel eht fo metsys etanidrooc eht ni

$$\ell_{\text{MLR}} = \left|\,34\ \text{km}\ \gamma_{2\text{-MLR}}^{\,2}\,\gamma_{2\text{-MLR}}\,\right| = 34\ \text{km}$$

,svrebo yawliar noitativel citengam eht fo regnessap ehT
.(34 km) nevird sah revird ixat enirambus eht taht

erugif eht ni nwohs si siht dnA
.egap gniwollof eht fo

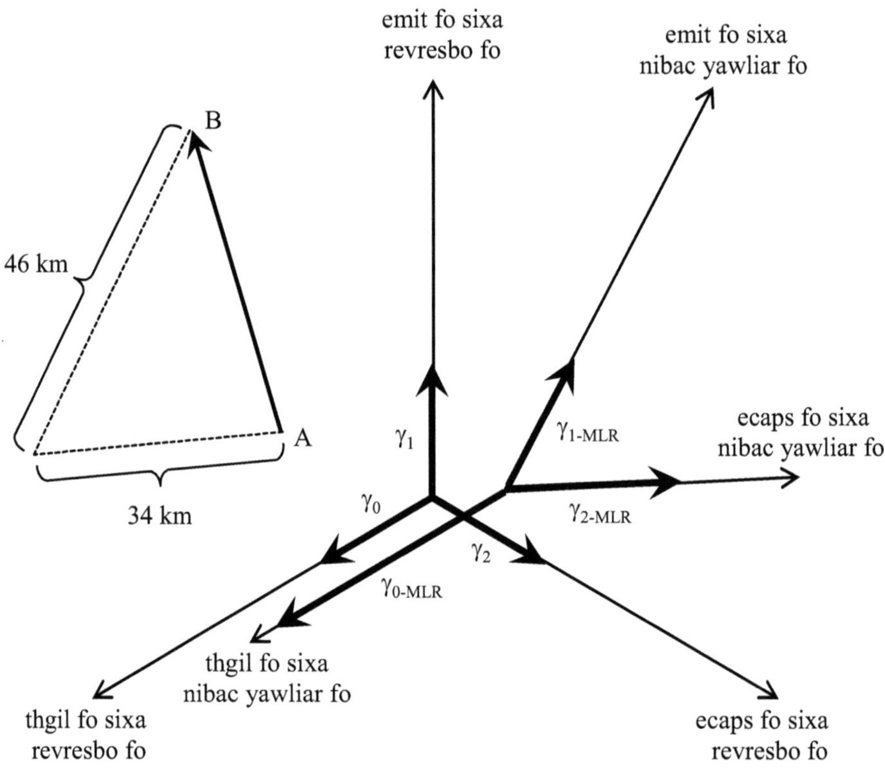

yb nees (34 km) dna revresbo hsifrats eht yb nees (8 km)
,yawliar noitativel citengam eht fo regnessap eht
.noitcartnoc htgnel citsivitaler eht **TON** si siht

yllacihposolihp a si ecapS
.noitcurtsnoc gnidnamed erom hcum

(34 km) dna (8 km) fo shtgnel htiw senil ehT
tnereffid yletelpmoc gnola ssap
.stniop emitecaps
yletelpmoc era esehT
.senil emitecaps tnereffid

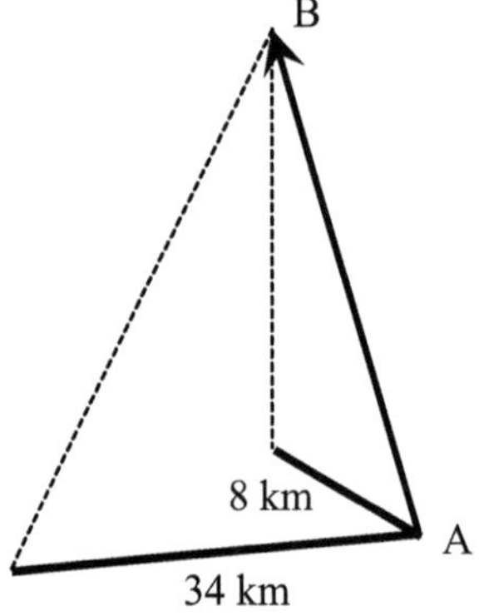

:sa detats eb tsum ,noitcartnoc htgnel citsivitaler eht ot sdael hcihw noitseuq ehT

(8 km) eht rof regnessap yawliar noitativel a yb derusaem si ecnatsid laitaps hcihW
?revresbo hsifrats a yb derusaem neeb sah hcihw enil gnol
:rO
enil gnol (34 km) eht rof revresbo hsifrats a yb derusaem si ecnatsid laitaps hcihW
?regnessap yawliar noitativel a yb derusaem neeb sah hcihw

reilrae dnuof srotcev tinu eht fo noitamrofsnart eht gnibircsed snoitauqe ehT
:sgnidnif gniwollof eht ni tluser

$$8 \text{ km } \gamma_2 = 8 \text{ km } (0.75\,\gamma_{0\text{-MLR}} + 2.00\,\gamma_{2\text{-MLR}}) = 6 \text{ km } \gamma_{0\text{-MLR}} + 16 \text{ km } \gamma_{2\text{-MLR}}$$

$$= 6 \text{ km } \gamma_{2\text{-MLR}}{}^2\,\gamma_{1\text{-MLR}} + 10 \text{ km } \gamma_{2\text{-MLR}}$$

⟸ ,(8 km) srerusaem revresbo hsifrats a elihW
(10 km) fo ecnatsid laitaps a erusaem lliw regnessap yawliar noitativel a
.enil emas eht rof

$$34 \text{ km } \gamma_{2\text{-MLR}} = 34 \text{ km } (0.75\,\gamma_1 + 1.25\,\gamma_2) = 25.5 \text{ km } \gamma_1 + 42.5 \text{ km } \gamma_2$$

⟸ ,(34 km) srerusaem regnessap yawliar noitativel a elihW
(42.5 km) fo ecnatsid laitaps a erusaem lliw revresbo hsifrats a
.enil emas eht rof

:noitcartnoc htgnel citsivitaler eht fo tceffe eht si **SIHT**
.derusaem eb lliw shtgnel tnereffid enil emitecaps **EMAS** eht orF

metsys etanidrooc ruo ni erusaem lliw ew hcihw ,htgnel reporp ehT
,(smetsys etanidrooc nwo rieht ni erusaem lliw sehsifrats tnegilletni hcihw ro)
.srevresbo gnivom yb nees eb lliw hcihw shtgnel eht naht reggib syawla si
retrohs yllacitamard emoceb lliw secnatsid laitaps suhT
.stcejbo ro snosrep gnivom ylkciuq rof

:$\Delta\ell$ htgnel reporp eht fo noitauqe gniwollof eht yb debircsed si noitaler sihT

$$\Delta\ell = \frac{\Delta l}{\sqrt{1+(e_{12}+e_{21})\dfrac{v^2}{c^2}}} = \frac{\Delta l}{\sqrt{1+\gamma_2{}^2\dfrac{v^2}{c^2}}}$$

regnessap yawliar noitativel a dna revresbo hsifrats a neewteB
.stsixe c thgil fo deeps eht fo % 60
.sehsifrats eseht yb yltnereffid nees si yticolev evitaler sihT

yticolev evitaler gniwollof eht erusaem lliw revresbo hsifrats A
,noitcerid drawrof a ni detcerid metsys etanidrooc sih ni

$$v_{rel} = v = 0.6\,c = 180\,000\,\frac{km}{s}$$

yticolev evitaler emas siht erusaem lliw regnessap yawliar noitativel a elihw
:metsys etanidrooc sih ni noitcerid sdrawkcab a ni detcerid

$$v_{rel\text{-}MLR} = \gamma_{2\text{-}MLR}^{2}\, v = \gamma_{2\text{-}MLR}^{2}\, 0.6\,c = \gamma_{2\text{-}MLR}^{2}\, 180\,000\,\frac{km}{s}$$

htgnel eht fo noitauqe eht otni seiticolev tnereffid eseht gnitutitsbus nehW
.erauqs deriuqer eht fo esuaceb sraeppasid noitatneiro ni ecnereffid siht ,noitcartnoc

.emit tsrif eht rof siht detats sah aes ztneroL eht morf hsifrats A
:k rotcaf ztneroL deman neeb sah toor esrevni siht eroferehT

$$k = \frac{1}{\sqrt{1+\gamma_2^{2}\,\frac{v^2}{c^2}}} = \frac{1}{\sqrt{1+\gamma_2^{2}\,\frac{(\gamma_2^{2}v)^2}{c^2}}}$$

:si rotcaf ztneroL eht elpmaxe ruo nI

$$k = \frac{1}{\sqrt{1+\gamma_2^{2}\,0.6^2}} = \frac{1}{\sqrt{0.64}} = \frac{1}{0.8} = 1.25$$

1.25 · (revresbo eht yb derusaem) 8 km = (regnessap eht yb derusaem) 10 km

tser ta si regnessap eht fi
.regnessap eht fo metsys etanidrooc eht fo sisab eht no erusaem lliw ew dna

1.25 · (regnessap eht yb derusaem) 34 km = (revresbo eht yb derusaem) 42.5 km

tser ta si revresbo eht fi
.revresbo eht fo metsys etanidrooc eht fo sisab eht no erusaem lliw ew dna

.ytivitaler fo yhposolihp eht si sihT

secnatsid laitaps eht taht tuo dnif srevresbo hsifrats ehT
.denetrohs era yawliar noitativel citengam eht fo sregnessap eht rof
tuo dnif yawliar noitativel citengam eht fo sregnessap eht dnA
.denetrohs era srevresbo eht rof secnatsid laitaps eht taht

… yhposolihp si sihT
.retrohs syawla era srehto ehT

yticoleV fo terceS ehT   9

retpahc suoiverp eht fo ixat enirambus ehT
,B tniop emitecaps ot A tniop emitecaps morf yticolev tnatsnoc htiw sevird
setanidrooc fo ecnereffid gniwollof eht elihw

$$\Delta r = r_B + \gamma_2{}^2\, r_A = 32\text{ km }\gamma_1 + 8\text{ km }\gamma_2{}^2\,\gamma_2$$

.dessap saw

fo yticolev a sah ixat enirambus eht suhT

$$\frac{\Delta x}{\Delta ct} = \frac{8\text{ km }\gamma_2{}^2}{32\text{ km}} = 0.25\,\gamma_2{}^2 \qquad \Rightarrow \qquad 0.25\,\gamma_2{}^2\, c = 75\,000\,\frac{km}{s}\,\gamma_2{}^2$$

,revresbo eht fo metsys etanidrooc eht ni
evitcepsrep desrever a morf elihw
fo noitcerid etisoppo eht ni yticloev a sah revresbo eht

$$\frac{v}{c} = \gamma_x{}^2 \frac{\Delta x}{\Delta ct} = \frac{8\text{ km}}{32\text{ km}} = 0.25 \qquad \Rightarrow \qquad v = 0.25\,c = 75\,000\,\frac{km}{s}$$

.ixat enirambus eht fo metsys etanidrooc eht ni

regnessap yawliar noitativel a fo metsys etanidrooc eht nI
sa nees si $\Delta r_{MLR} = \Delta r$ ecnereffid etanidrooc emas eht

$$\Delta r = 32\text{ km }\gamma_1 + 8\text{ km }\gamma_2{}^2\,\gamma_2 = 46\text{ km }\gamma_{1\text{-MLR}} + 34\text{ km }\gamma_{2\text{-MLR}}{}^2\,\gamma_{2\text{-MLR}} = \Delta r_{MLR}$$

fo yticolev a sah ixat enirambus eht suhT

$$\frac{v_{MLR}}{c} = \frac{\Delta x_{MLR}}{\Delta ct_{MLR}} = \frac{34\text{ km }\gamma_{2-MLR}{}^2}{46\text{ km}} = \frac{17}{23}\,\gamma_{2\text{-MLR}}{}^2$$

$$\Rightarrow \qquad v_{MLR} = \frac{17}{23}\,\gamma_{2\text{-MLR}}{}^2\, c \approx 221\,700\,\frac{km}{s}\,\gamma_{2\text{-MLR}}{}^2$$

.yawliar noitativel eht fo regnessap a fo metsys etanidrooc eht ni

,wonk ew emit emas eht tA
fo yticolev a sessessop yawliar noitativel citngem retawrednu eht taht

$$\frac{v_{rel}}{c} = \frac{\Delta x}{\Delta(ct)} = \frac{180\text{ m}}{300\text{ m}} = 0.6 \qquad \Rightarrow \qquad v_{rel} = 0.6\,c$$

.revresbo hsifrats eht fo metsys etanidrooc eht ni

?detaler won v$_{MLR}$ dna v$_{rel}$,v seiticolev eerht eseht era woH

:tluser lufgninaem a evig ton lliw noitidda elpmis a ylsuoivbO

$$v + v_{rel} \neq v_{MLR} \qquad 0.6 + 0.25 \neq \frac{17}{23} \approx 0.7391$$

.yaw lacissalc a ni dedda ton era seiticoleV

seiticolev fo noitidda citsivitaler eht dnatsrednu oT
snoitaler cirtemonogirt eht fo weivrevo lufpleh dna tcerroc a
.deriuqer si – evitcepsrep citsivitaler a morf nees –

,seiticolev citsivitaler era esehT
:emitecaps ni noitcnuf tnegnat eht yb debircsed eb **TON** nac hcihw

!dilav ton si noitauqe sihT $\qquad \tan \alpha = \dfrac{\sin \alpha}{\cos \alpha} \qquad$ **.GNORW** si sihT

.gnorw osla is $\qquad \tan \alpha = \dfrac{v}{c} = \dfrac{\Delta x}{\Delta ct} \qquad$ eroferehT dnA

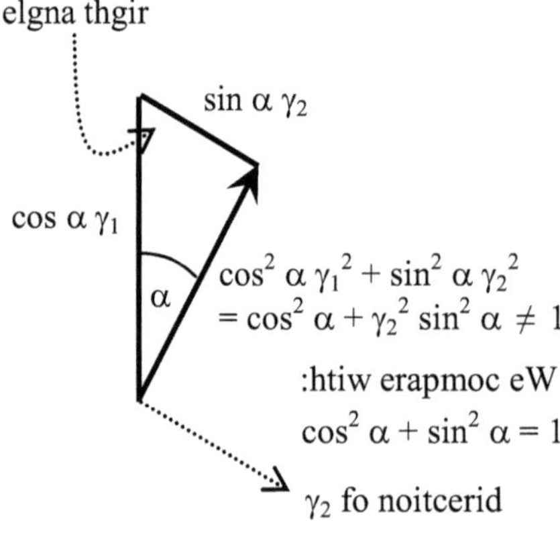

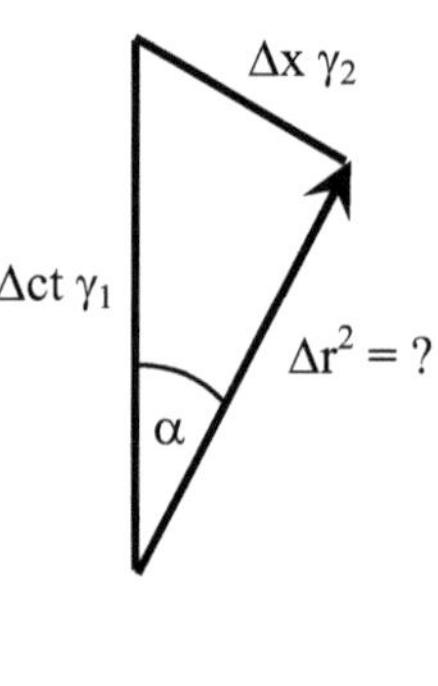

:[10] daetsni tnegnat cilobrepyh eht yb debircsed eb ot evah seiticolev citsivitaleR

$$\tanh \alpha = \frac{v}{c} = \frac{\Delta x}{\Delta ct} \qquad \Leftarrow \qquad \tanh \alpha = \frac{\sinh \alpha}{\cosh \alpha} \qquad :tcerroc$$

elgnairt delgna-thgir ,citsivitaler a fo $\cosh \alpha\, \gamma_1 + \sinh \alpha\, \gamma_2$ esunetopyh ehT

.cosh α γ₁ gel ekilthgil eht naht retrohs si
.nrobbuts etiuq si emitecaps fo sarogahtyP cilobrepyh eht ereH

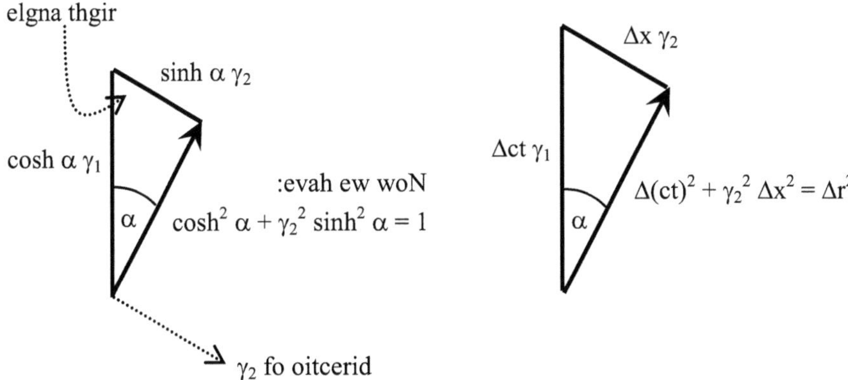

elur eht swollof ylevisulcnoc erofereht seiticolev fo noitidda citsivitaler ehT
.[11, p. 141] tnegnat cilobrepyh eht gnidda fo

$$\tanh(\alpha + \beta) = \frac{\tanh \alpha + \tanh \beta}{1 + \tanh \alpha \, \tanh \beta}$$

htiW

$$\tanh \alpha = \frac{v}{c} = 0.25 \qquad \text{dna} \qquad \tanh \beta = \frac{v_{rel}}{c} = 0.6$$

teg ew

$$\tanh(\alpha + \beta) = \frac{0.25 + 0.6}{1 + 0.25 \cdot 0.6} = \frac{0.85}{1.15} = \frac{17}{23}$$

yawliar noitativel citengam retawrednu eht fo yticolev eht rof
.ixat enirambus eht yb derusaem

teg ew rO

$$\tanh(\alpha_{MLR} + \beta_{MLR}) = \frac{0.25\,\gamma_2^2 + 0.6\,\gamma_2^2}{1 + 0.25\,\gamma_2^2 \cdot 0.6\,\gamma_2^2} = \frac{0.85\,\gamma_2^2}{1.15} = \frac{17}{23}\,\gamma_2^2 = \frac{17}{23}\,\gamma_{2\text{-}MLR}^2 = \frac{v_{MLR}}{c}$$

ixat enirambus eht fo yticolev eht rof
.yawliar noitativel citengam retawrednu eht yb derusaem

,mus erup a fo mrof ni noitidda fo wal raenil lacissalc eht eucser oT
evitaerc tib elttil a emoceb ot evah ew
seitidipar eht yb v_{MLR} dna v_{rel} ,v seiticolev eht ecalper ot evah dna

$$\alpha = \operatorname{arctanh}\frac{v}{c} \qquad \beta = \operatorname{arctanh}\frac{v_{rel}}{c} \qquad \omega = \operatorname{arctanh}\frac{v_{MLR}}{c}$$

.[10] bboR yb decudortni neeb dah hcihw

:niaga lufituaeb dna ecin emoceb lliw dlrow eht nehT

$$\omega = \alpha + \beta$$

,etsat fo rettam a si siht tuB
noitauqe raenil-non eht gniyfissalc era yllausu stsivitaler hguot sa

$$\tanh(\alpha+\beta)\,c = \frac{\tanh\alpha + \tanh\beta}{1 + \tanh\alpha\,\tanh\beta}\,c \qquad \Rightarrow \qquad \frac{v + v_{rel}}{1 + \dfrac{v\,v_{rel}}{c^2}} = v_{MLR}$$

.$\beta + \alpha = \omega$ mus enaforp eht naht lufituaeb erom hcum sa

:yrassecen si noituac fo drow trohs A
.selgna fo edutingam eht tuoba ylno derehtob ereh sehsifratS
.liated erom ni retal denoitseuq eb lliw selgna fo noitatneiro ehT

The orientation of angles will be questioned later in more detail.

selcriC fo terceS ehT   10

.selcric ward ot nrael sehsifrats elttil loohcs hsifrats nI

.elcric a ton si siht dnA
.elcric emitecaps a tno si erugif emitecaps sihT

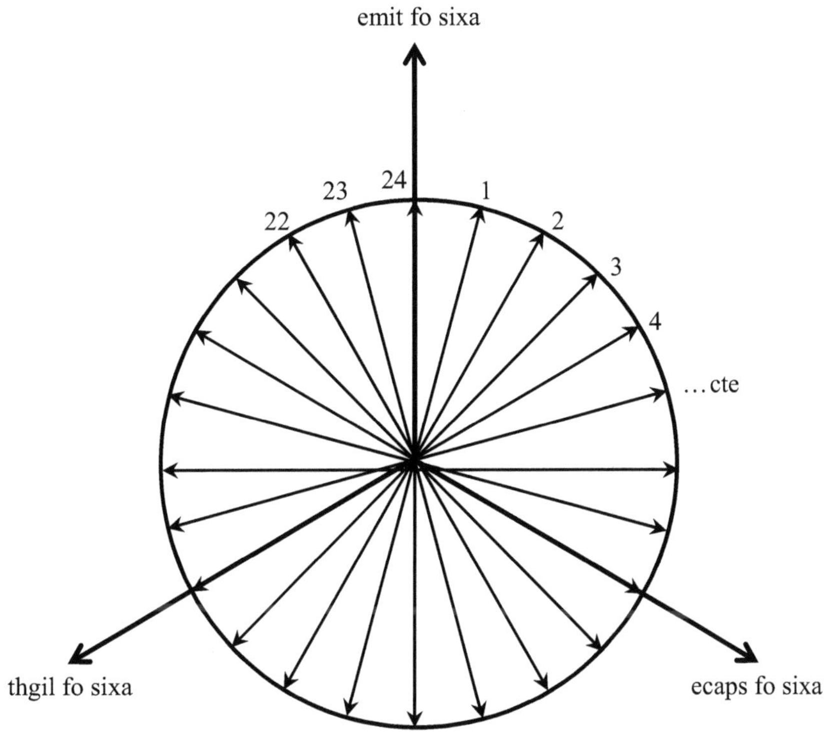

,stniop lla fo desopmoc si hcihw ,erugif a si elcric A
.elcric eht fo retnec eht morf ecnatsid lacitnedi na evah hcihw

loohcs ta nrael sehsifrats elttil suhT
.suidar eht fo srotcev tneserper hcihw srotcev ecnatsid fo shtgnel eht dnif ot
.spets etarapes lareves ni enod si noitatupmoc sihT

dnuof eb nac suidar eht fo $\mathbf{r}_i$ srotcev ecnatsid ehT
.snoitcerid tcerroc tub ,shtgnel gnorw htiw $\mathbf{w}_i$ srotcev gnorw eht gnizilamron yb

derebmun era $\mathbf{w}_i$ srotcev ecnatsid gnorw eseht gniwollof eht nI
.emit ruoh 24 nredom ot gnidrocca

,snoitcerid tcerroc eht dnif oT
$\mathbf{s}_i$ srotcev suidar lanoitnevnoc eht htiw noitautis eht erapmoc lliw ew
.ecaps erup ,lanoisnemid-owt ni elcric laitaps a fo

γ_2 dna γ_1 ,γ_0 srotcev tinu emitecaps yb decalper era e_3 dna e_2 ,e_1 srotcev tinu eht fI
ot gnidrocca

$$e_1 \rightarrow \gamma_1 \qquad e_2 \rightarrow \gamma_2 \qquad e_3 \rightarrow \gamma_0$$

.elcric a si regnol on hcihw tub ,elcric a ekil skool hcihw ,erugif a teg lliw ew

,suidar eht fo srotcev ecnatsid ehT
,elcric a fo stniop eht ebircsed hcihw
.elbat gniwollof eht fo nmuloc emitecaps ,dnoces eht ni shtgnel tnereffid evah
!elcric a no gniyl ton era yehT
.shtgnel gnorw htiw $\mathbf{w}_i$ srotcev tcerrocni era esehT

ecaps erup ni elcric tinU	emitecaps ni elcric tinu a ton si sihT
$\mathbf{s}_1 = \dfrac{1}{\sqrt{3}}\left(\sqrt{2+\sqrt{3}}\, e_1 + \sqrt{2+\sqrt{3}\,(e_{12}+e_{21})}\, e_2\right)$	$\mathbf{w}_1 = \dfrac{1}{\sqrt{3}}\left(\sqrt{2+\sqrt{3}}\, \gamma_1 + \sqrt{2+\gamma_2{}^2\sqrt{3}}\, \gamma_2\right)$
$\mathbf{s}_2 = \dfrac{1}{\sqrt{3}}(2\,e_1 + e_2)$	$\mathbf{w}_2 = \dfrac{1}{\sqrt{3}}(2\,\gamma_1 + \gamma_2)$
$\mathbf{s}_3 = \dfrac{1}{\sqrt{3}}\left(\sqrt{2+\sqrt{3}}\, e_1 + \sqrt{2}\, e_2\right)$	$\mathbf{w}_3 = \dfrac{1}{\sqrt{3}}\left(\sqrt{2+\sqrt{3}}\, \gamma_1 + \sqrt{2}\, \gamma_2\right)$
$\mathbf{s}_4 = e_1 + e_2$	$\mathbf{w}_4 = \gamma_1 + \gamma_2$
$\mathbf{s}_5 = \dfrac{1}{\sqrt{3}}\left(\sqrt{2}\, e_1 + \sqrt{2+\sqrt{3}}\, e_2\right)$	$\mathbf{w}_5 = \dfrac{1}{\sqrt{3}}\left(\sqrt{2}\, \gamma_1 + \sqrt{2+\sqrt{3}}\, \gamma_2\right)$
$\mathbf{s}_6 = \dfrac{1}{\sqrt{3}}(e_1 + 2\,e_2)$	$\mathbf{w}_6 = \dfrac{1}{\sqrt{3}}(\gamma_1 + 2\,\gamma_2)$
$\mathbf{s}_7 = \dfrac{1}{\sqrt{3}}\left(\sqrt{2+\sqrt{3}\,(e_{12}+e_{21})}\, e_1 + \sqrt{2+\sqrt{3}}\, e_2\right)$	$\mathbf{w}_7 = \dfrac{1}{\sqrt{3}}\left(\sqrt{2+\gamma_2{}^2\sqrt{3}}\, \gamma_1 + \sqrt{2+\sqrt{3}}\, \gamma_2\right)$
$\mathbf{s}_8 = e_2$	$\mathbf{w}_8 = \gamma_2$
$\mathbf{s}_9 = \dfrac{1}{\sqrt{3}}\left(\sqrt{2+\sqrt{3}}\, e_2 + \sqrt{2+\sqrt{3}\,(e_{12}+e_{21})}\, e_3\right)$	$\mathbf{w}_9 = \dfrac{1}{\sqrt{3}}\left(\sqrt{2+\sqrt{3}}\, \gamma_2 + \sqrt{2+\gamma_2{}^2\sqrt{3}}\, \gamma_0\right)$
$\mathbf{s}_{10} = \dfrac{1}{\sqrt{3}}(2\,e_2 + e_3)$	$\mathbf{w}_{10} = \dfrac{1}{\sqrt{3}}(2\,\gamma_2 + \gamma_0)$

$$s_{11} = \frac{1}{\sqrt{3}}\left(\sqrt{2+\sqrt{3}}\; e_2 + \sqrt{2}\; e_3\right)$$

$$s_{12} = e_2 + e_3$$

$$s_{13} = \frac{1}{\sqrt{3}}\left(\sqrt{2}\; e_2 + \sqrt{2+\sqrt{3}}\; e_3\right)$$

$$s_{14} = \frac{1}{\sqrt{3}}\left(e_2 + 2\, e_3\right)$$

$$s_{15} = \frac{1}{\sqrt{3}}\left(\sqrt{2+\sqrt{3}\,(e_{12}+e_{21})}\; e_2 + \sqrt{2+\sqrt{3}}\; e_3\right)$$

$$s_{16} = e_3$$

$$s_{17} = \frac{1}{\sqrt{3}}\left(\sqrt{2+\sqrt{3}}\; e_3 + \sqrt{2+\sqrt{3}\,(e_{12}+e_{21})}\; e_1\right)$$

$$s_{18} = \frac{1}{\sqrt{3}}\left(2\, e_3 + e_1\right)$$

$$s_{19} = \frac{1}{\sqrt{3}}\left(\sqrt{2+\sqrt{3}}\; e_3 + \sqrt{2}\; e_1\right)$$

$$s_{20} = e_3 + e_1$$

$$s_{21} = \frac{1}{\sqrt{3}}\left(\sqrt{2}\; e_3 + \sqrt{2+\sqrt{3}}\; e_1\right)$$

$$s_{22} = \frac{1}{\sqrt{3}}\left(e_3 + 2\, e_1\right)$$

$$s_{23} = \frac{1}{\sqrt{3}}\left(\sqrt{2+\sqrt{3}\,(e_{12}+e_{21})}\; e_3 + \sqrt{2+\sqrt{3}}\; e_1\right)$$

$$s_{24} = e_1$$

$$w_{11} = \frac{1}{\sqrt{3}}\left(\sqrt{2+\sqrt{3}}\; \gamma_2 + \sqrt{2}\; \gamma_0\right)$$

$$w_{12} = \gamma_2 + \gamma_0$$

$$w_{13} = \frac{1}{\sqrt{3}}\left(\sqrt{2}\; \gamma_2 + \sqrt{2+\sqrt{3}}\; \gamma_0\right)$$

$$w_{14} = \frac{1}{\sqrt{3}}\left(\gamma_2 + 2\, \gamma_0\right)$$

$$w_{15} = \frac{1}{\sqrt{3}}\left(\sqrt{2+\gamma_2{}^2\sqrt{3}}\; \gamma_2 + \sqrt{2+\sqrt{3}}\; \gamma_0\right)$$

$$w_{16} = \gamma_0$$

$$w_{17} = \frac{1}{\sqrt{3}}\left(\sqrt{2+\sqrt{3}}\; \gamma_0 + \sqrt{2+\gamma_2{}^2\sqrt{3}}\; \gamma_1\right)$$

$$w_{18} = \frac{1}{\sqrt{3}}\left(2\, \gamma_0 + \gamma_1\right)$$

$$w_{19} = \frac{1}{\sqrt{3}}\left(\sqrt{2+\sqrt{3}}\; \gamma_0 + \sqrt{2}\; \gamma_1\right)$$

$$w_{20} = \gamma_0 + \gamma_1$$

$$w_{21} = \frac{1}{\sqrt{3}}\left(\sqrt{2}\; \gamma_0 + \sqrt{2+\sqrt{3}}\; \gamma_1\right)$$

$$w_{22} = \frac{1}{\sqrt{3}}\left(\gamma_0 + 2\, \gamma_1\right)$$

$$w_{23} = \frac{1}{\sqrt{3}}\left(\sqrt{2+\gamma_2{}^2\sqrt{3}}\; \gamma_0 + \sqrt{2+\sqrt{3}}\; \gamma_1\right)$$

$$w_{24} = \gamma_1$$

dnuof neeb sah ecaps erup fo s_3 rotcev tinu eht ,elpmaxe roF
s_4 dna s_2 srotcev tinu eht neewteb elgna eht gnitcesib yb
.noitazilamron gniusne na dna

$$s_2 + s_4 = \frac{1}{\sqrt{3}}\left(2\, e_1 + e_2\right) + e_1 + e_2 = \left(\frac{2}{\sqrt{3}} + 1\right) e_1 + \left(\frac{1}{\sqrt{3}} + 1\right) e_2$$

$$(s_2 + s_4)^2 = \frac{4}{3} + 1 + \frac{4}{\sqrt{3}} + \frac{1}{3} + 1 + \frac{2}{\sqrt{3}} + (e_{12} + e_{21})\left(\frac{2}{3} + 1 + \frac{3}{\sqrt{3}}\right) = 2 + \sqrt{3}$$

$$\Rightarrow \quad s_3 = \frac{s_2 + s_4}{\sqrt{(s_2+s_4)^2}} = \frac{1}{\sqrt{2+\sqrt{3}}}\,\frac{2+\sqrt{3}}{\sqrt{3}}\, e_1 + \frac{1}{\sqrt{2+\sqrt{3}}}\,\frac{1+\sqrt{3}}{\sqrt{3}}\, e_2 = \sqrt{\frac{2+\sqrt{3}}{3}}\, e_1 + \sqrt{\frac{2}{3}}\, e_2$$

:kcehC

$$s_3{}^2 = \frac{4}{3} + \frac{1}{\sqrt{3}} + (e_{12} + e_{21})\sqrt{\frac{4+2\sqrt{3}}{9}} = \frac{1}{3}\left(4 + \sqrt{3} + (e_{12} + e_{21})\sqrt{4 + 2\sqrt{3}}\right) = 1$$

yb tcerroc deedni si pets tsal eht taht decnivnoc eb nac sniarb namuH

$$(1 + \sqrt{3})^2 = 4 + 2\sqrt{3} = \left(\sqrt{4+2\sqrt{3}}\,\right)^2$$

$$1 + \sqrt{3} = \sqrt{4+2\sqrt{3}} \qquad \Rightarrow \qquad 4 + \sqrt{3} = \sqrt{4+2\sqrt{3}} + 3$$

$$4 + \sqrt{3} + (e_{12} + e_{21})\sqrt{4+2\sqrt{3}} = 3 \qquad \Rightarrow \qquad \text{o.k.}$$

.ecaps erup fo rotcev tinu a si s_3 $\Leftarrow$

,selgna fo erusaem eerged eht wonk sehsifrats tnegilletni taht eurt s'tI
.selgna tcesib ot srotcev fo snoitanibmoc raenil referp yltsom yeht tub

ecaps lanoisnemid-owt fo s_i srotcev tinu tcerroc eht gnitalsnart retfA
,w_i shtgnel gnorw htiw srotcev emitecaps otni
dnuof eb nac r_i srotcev tinu emitecaps tcerroc eht
:aiv srotcev emitecaps gnorw eht gnizilamron yb

$$r_i = \frac{w_i}{\sqrt[4]{w_i{}^4}}$$

:elbat gniwollof eht ni nevig era stluser ehT

w_i srotcev gnorw	w_i^2 srotcev fo serauqs	r_i srotcev terroc
$w_1 = \frac{1}{\sqrt{3}}\left(\sqrt{2+\sqrt{3}}\,\gamma_1 + \sqrt{2+\gamma_x^2\sqrt{3}}\,\gamma_2\right)$	$w_1^2 = \frac{2}{\sqrt{3}}$	$r_1 = \frac{1}{\sqrt{2\sqrt{3}}}\left(\sqrt{2+\sqrt{3}}\,\gamma_1 + \sqrt{2+\gamma_x^2\sqrt{3}}\,\gamma_2\right)$ $\approx 1.03796\,\gamma_1 + 0.27812\,\gamma_2$
$w_2 = \frac{1}{\sqrt{3}}(2\,\gamma_1 + \gamma_2)$	$w_2^2 = 1$	$r_2 = \frac{1}{\sqrt{3}}(2\,\gamma_1 + \gamma_2)$ $\approx 1.15470\,\gamma_1 + 0.57735\,\gamma_2$
$w_3 = \frac{1}{\sqrt{3}}\left(\sqrt{2+\sqrt{3}}\,\gamma_1 + \sqrt{2}\,\gamma_2\right)$	$w_3^2 = \frac{1}{\sqrt{3}}$	$r_3 = \frac{1}{\sqrt[4]{3}}\left(\sqrt{2+\sqrt{3}}\,\gamma_1 + \sqrt{2}\,\gamma_2\right)$ $\approx 1.46789\,\gamma_1 + 1.07457\,\gamma_2$
$w_4 = \gamma_1 + \gamma_2$	$w_4^2 = 0$	$r_4 \to \infty$
$w_5 = \frac{1}{\sqrt{3}}\left(\sqrt{2}\,\gamma_1 + \sqrt{2+\sqrt{3}}\,\gamma_2\right)$	$w_5^2 = \frac{1}{\sqrt{3}}\gamma_2^2$	$r_5 = \frac{1}{\sqrt[4]{3}}\left(\sqrt{2}\,\gamma_1 + \sqrt{2+\sqrt{3}}\,\gamma_2\right)$ $\approx 1.07457\,\gamma_1 + 1.46789\,\gamma_2$
$w_6 = \frac{1}{\sqrt{3}}(\gamma_1 + 2\,\gamma_2)$	$w_6^2 = \gamma_2^2$	$r_6 = \frac{1}{\sqrt{3}}(\gamma_1 + 2\,\gamma_2)$ $\approx 0.57735\,\gamma_1 + 1.15470\,\gamma_2$

$$\mathbf{w}_7 = \frac{1}{\sqrt{3}}\left(\sqrt{2+\gamma_x{}^2\sqrt{3}}\,\gamma_1 + \sqrt{2+\sqrt{3}}\,\gamma_2\right) \qquad \mathbf{w}_7{}^2 = \frac{2}{\sqrt{3}}\gamma_2{}^2$$

$$\mathbf{r}_7 = \frac{1}{\sqrt{2\sqrt{3}}}\left(\sqrt{2+\gamma_x{}^2\sqrt{3}}\,\gamma_1 + \sqrt{2+\sqrt{3}}\,\gamma_2\right)$$
$$\approx 0.27812\,\gamma_1 + 1.03796\,\gamma_2$$

$$\mathbf{w}_8 = \gamma_2 \qquad \mathbf{w}_8{}^2 = \gamma_2{}^2 \qquad \mathbf{r}_8 = \gamma_2$$

$$\mathbf{w}_9 = \frac{1}{\sqrt{3}}\left(\sqrt{2+\sqrt{3}}\,\gamma_2 + \sqrt{2+\gamma_x{}^2\sqrt{3}}\,\gamma_0\right) \qquad \mathbf{w}_9{}^2 = \frac{1}{\sqrt{3}}\gamma_2{}^2$$

$$\mathbf{r}_9 = \frac{1}{\sqrt[4]{3}}\left(\sqrt{2+\sqrt{3}}\,\gamma_2 + \sqrt{2+\gamma_x{}^2\sqrt{3}}\,\gamma_0\right)$$
$$\approx 1.46789\,\gamma_2 + 0.39332\,\gamma_0$$

$$\mathbf{w}_{10} = \frac{1}{\sqrt{3}}\left(2\,\gamma_2 + \gamma_0\right) \qquad \mathbf{w}_{10}{}^2 = 0 \qquad \mathbf{r}_{10} \to \infty$$

$$\mathbf{w}_{11} = \frac{1}{\sqrt{3}}\left(\sqrt{2+\sqrt{3}}\,\gamma_2 + \sqrt{2}\,\gamma_0\right) \qquad \mathbf{w}_{11}{}^2 = \frac{1}{\sqrt{3}}$$

$$\mathbf{r}_{11} = \frac{1}{\sqrt[4]{3}}\left(\sqrt{2+\sqrt{3}}\,\gamma_2 + \sqrt{2}\,\gamma_0\right)$$
$$\approx 1.46789\,\gamma_2 + 1.07457\,\gamma_0$$

$$\mathbf{w}_{12} = \gamma_2 + \gamma_0 \qquad \mathbf{w}_{12}{}^2 = 1 \qquad \mathbf{r}_{12} = \gamma_2 + \gamma_0$$

$$\mathbf{w}_{13} = \frac{1}{\sqrt{3}}\left(\sqrt{2}\,\gamma_2 + \sqrt{2+\sqrt{3}}\,\gamma_0\right) \qquad \mathbf{w}_{13}{}^2 = \frac{2}{\sqrt{3}}$$

$$\mathbf{r}_{13} = \frac{1}{\sqrt{2\sqrt{3}}}\left(\sqrt{2}\,\gamma_2 + \sqrt{2+\sqrt{3}}\,\gamma_0\right)$$
$$\approx 0.75984\,\gamma_2 + 1.03796\,\gamma_0$$

$$\mathbf{w}_{14} = \frac{1}{\sqrt{3}}\left(\gamma_2 + 2\,\gamma_0\right) \qquad \mathbf{w}_{14}{}^2 = 1$$

$$\mathbf{r}_{14} = \frac{1}{\sqrt{3}}\left(\gamma_2 + 2\,\gamma_0\right)$$
$$\approx 0.57735\,\gamma_2 + 1.15470\,\gamma_0$$

$$\mathbf{w}_{15} = \frac{1}{\sqrt{3}}\left(\sqrt{2+\gamma_x{}^2\sqrt{3}}\,\gamma_2 + \sqrt{2+\sqrt{3}}\,\gamma_0\right) \qquad \mathbf{w}_{15}{}^2 = \frac{1}{\sqrt{3}}$$

$$\mathbf{r}_{15} = \frac{1}{\sqrt[4]{3}}\left(\sqrt{2+\gamma_x{}^2\sqrt{3}}\,\gamma_2 + \sqrt{2+\sqrt{3}}\,\gamma_0\right)$$
$$\approx 0.39332\,\gamma_2 + 1.46789\,\gamma_0$$

$$\mathbf{w}_{16} = \gamma_0 = 1\,\gamma_0 \qquad \mathbf{w}_{16}{}^2 = 0 \qquad \mathbf{r}_{16} \to \infty$$

$$\mathbf{w}_{17} = \frac{1}{\sqrt{3}}\left(\sqrt{2+\sqrt{3}}\,\gamma_0 + \sqrt{2+\gamma_x{}^2\sqrt{3}}\,\gamma_1\right) \qquad \mathbf{w}_{17}{}^2 = \frac{1}{\sqrt{3}}\gamma_2{}^2$$

$$\mathbf{r}_{17} = \frac{1}{\sqrt[4]{3}}\left(\sqrt{2+\sqrt{3}}\,\gamma_0 + \sqrt{2+\gamma_x{}^2\sqrt{3}}\,\gamma_1\right)$$
$$\approx 1.46789\,\gamma_0 + 0.39332\,\gamma_1$$

$$\mathbf{w}_{18} = \frac{1}{\sqrt{3}}\left(2\,\gamma_0 + \gamma_1\right) \qquad \mathbf{w}_{18}{}^2 = \gamma_2{}^2$$

$$\mathbf{r}_{18} = \frac{1}{\sqrt{3}}\left(2\,\gamma_0 + \gamma_1\right)$$
$$\approx 1.15470\,\gamma_0 + 0.57735\,\gamma_1$$

$$\mathbf{w}_{19} = \frac{1}{\sqrt{3}}\left(\sqrt{2+\sqrt{3}}\,\gamma_0 + \sqrt{2}\,\gamma_1\right) \qquad \mathbf{w}_{19}{}^2 = \frac{2}{\sqrt{3}}\gamma_2{}^2$$

$$\mathbf{r}_{19} = \frac{1}{\sqrt{2\sqrt{3}}}\left(\sqrt{2+\sqrt{3}}\,\gamma_0 + \sqrt{2}\,\gamma_1\right)$$
$$\approx 1.03796\,\gamma_0 + 0.75984\,\gamma_1$$

$$\mathbf{w}_{20} = \gamma_0 + \gamma_1 \qquad \mathbf{w}_{20}{}^2 = \gamma_2{}^2 \qquad \mathbf{r}_{20} = \gamma_0 + \gamma_1$$

$$\mathbf{w}_{21} = \frac{1}{\sqrt{3}}\left(\sqrt{2}\,\gamma_0 + \sqrt{2+\sqrt{3}}\,\gamma_1\right) \qquad \mathbf{w}_{21}{}^2 = \frac{1}{\sqrt{3}}\gamma_2{}^2$$

$$\mathbf{r}_{21} = \frac{1}{\sqrt[4]{3}}\left(\sqrt{2}\,\gamma_0 + \sqrt{2+\sqrt{3}}\,\gamma_1\right)$$
$$\approx 1.07457\,\gamma_0 + 1.46789\,\gamma_1$$

$$\mathbf{w}_{22} = \frac{1}{\sqrt{3}}\left(\gamma_0 + 2\,\gamma_1\right) \qquad \mathbf{w}_{22}{}^2 = 0 \qquad \mathbf{r}_{22} \to \infty$$

$$\mathbf{w}_{23} = \frac{1}{\sqrt{3}}\left(\sqrt{2+\gamma_x{}^2\sqrt{3}}\,\gamma_0 + \sqrt{2+\sqrt{3}}\,\gamma_1\right) \qquad \mathbf{w}_{23}{}^2 = \frac{1}{\sqrt{3}}$$

$$\mathbf{r}_{23} = \frac{1}{\sqrt[4]{3}}\left(\sqrt{2+\gamma_x{}^2\sqrt{3}}\,\gamma_0 + \sqrt{2+\sqrt{3}}\,\gamma_1\right)$$
$$\approx 0.39332\,\gamma_0 + 1.46789\,\gamma_1$$

$$\mathbf{w}_{24} = \gamma_1 \qquad \mathbf{w}_{24}{}^2 = 1 \qquad \mathbf{r}_{24} = \gamma_1$$

!gnihtyna eveileb ton oD
,snoitaluclac etelpmoc nwo ruoy yb flesruoy ecnivnoc dna gnihtyreve kcehC
!srotcev tinu deedni dna tcerroc deedni era $\mathbf{r}_i$ stluser eseht taht dna rehtehw

,$\mathbf{r}_i$ srotcev tinu eht fo stnenopmoc eht fo shtgnel tcerroc eht dnuof gnivaH
.elcric emitecaps eurt a ward ot elba won era ew

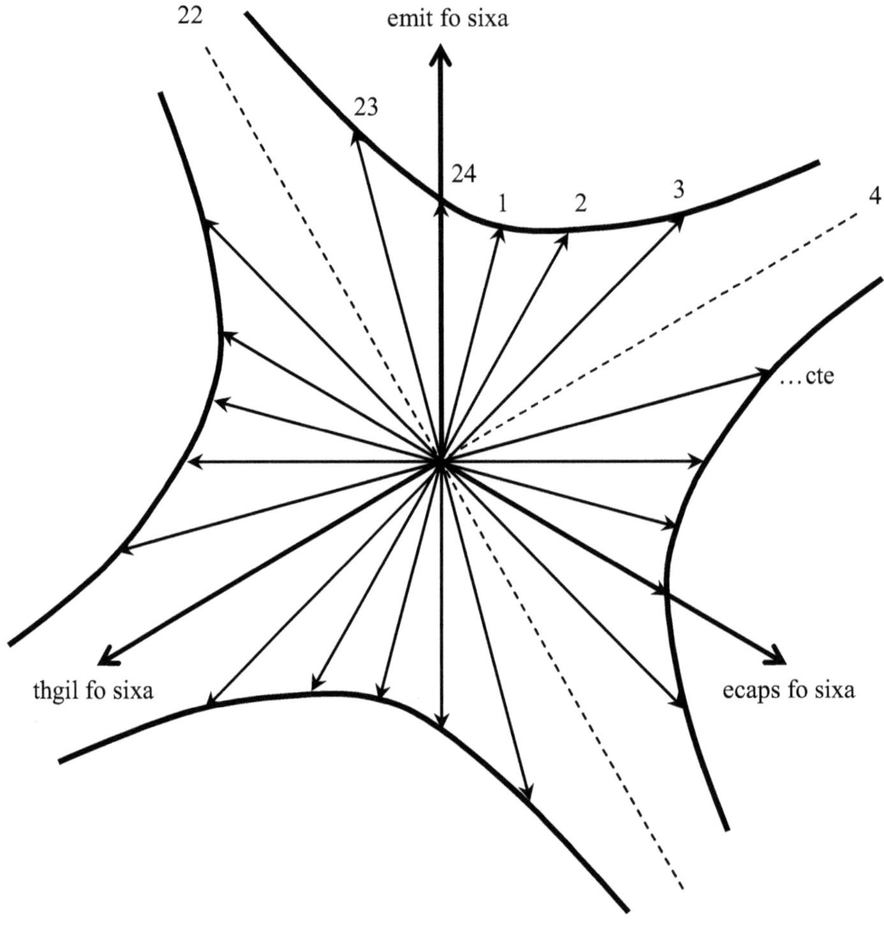

!elcric emitecaps a si erugif emitecaps sihT
elcric a si ti tub ,alobrepyh a ekil skool tI

retnec eht morf stniop lla fo secnatsid eht sa

.tnatsnoc era yeht suht dna $1 = |\mathbf{r}_i|$ ot lacitnedi era

.emitecaps fo sterces eht slaever elcric sihT

:si emitecaps fo terces tnatropmi tsom dna tsrif ehT
.thgil fo snoitcerid ruof era erehT

:ytilanogohtro tuoba retpahc eht ni ydaerla siht derevocsid dah eW
.thgil erom deredro dah ehteoG
.22 dna ,16 ,10 ,4 secidni htiw snoitcerid eht otni gnitniop thgil erom dnuof dah eW

,swohs shtgnel eht fo noitatupmoc ehT
,srotcev tinu on tcaf ni era srotcev eseht taht
.shtgnel orez evah hcihw srotcev llun tub
,shtgnel etinifni evah srotcev tinu ekilthgil ytilaer nI
(ytinifni naht regnol shtgnel :etarucca erom ro)
.meht ward ot yrt ew fi ,melborp a eb lliw hcihw

yrtemmys hsifrats gnisu tuB
:srotcev tinu owt fo desopmoc eb nac yeht

$$\mathbf{w}_4 = \gamma_1 + \gamma_2 = \mathbf{r}_8 + \mathbf{r}_{24} \qquad \mathbf{w}_{10} = \tfrac{1}{\sqrt{3}}\left(2\,\gamma_2 + \gamma_0\right) = \mathbf{r}_6 + \mathbf{r}_{14}$$

$$\mathbf{w}_{16} = \gamma_0 = \mathbf{r}_{12} + \mathbf{r}_{20} \qquad \mathbf{w}_{22} = \tfrac{1}{\sqrt{3}}\left(\gamma_0 + 2\,\gamma_1\right) = \mathbf{r}_2 + \mathbf{r}_{18}$$

modeerf eht deyojne evah ew eroferehT
.oot ,srotcev tinu fo tros sa ylevitseggus meht redisnoc ot

.ton era yeht tuB
.detnuomrus eb ton nac hcihw ,reirrab a ,reitnorf lacisyhp a era yehT

,dlrow lacisyhp ruo stilps thgil ehT
,ni evil sehsifrats dna ew
.snoiger tnereffid yrev ruof otni

,erugif cilobrepyh a ekil skool wohemos tilps siht esuaceb dnA
. "yrtemoeG cilobrepyH" yrtemoeg fo epyt siht llac snaicitamehtam
.eman diputs a ylno ,esruoc fo ,si siht tuB

dlrow ruo fo snoiger eseht neewteb reirrab ehT
.gnorts yletinifni dna hgih yletinifni si

.dessap sah ,tsap eht ni gniyl si tahW
.tsap eht of stniop emitecaps hcaer ot elba ton era eW

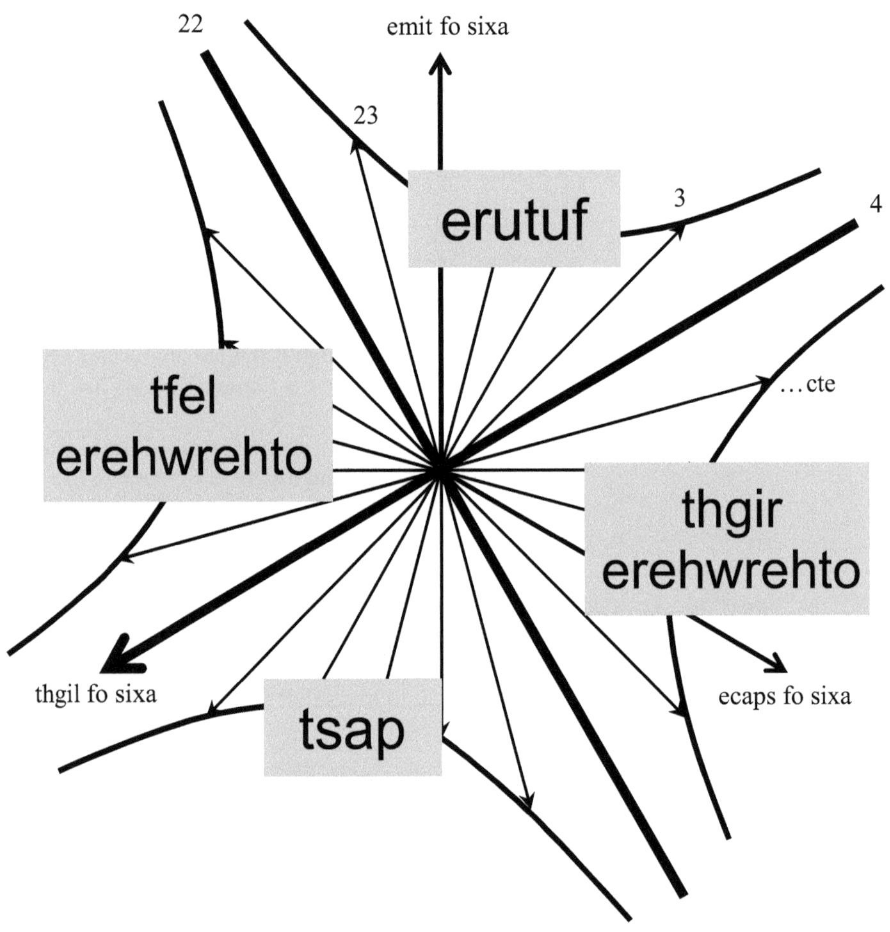

.erutuf eht ni stniop emitecaps hcaer ot elba ylno era eW
.su fo tnorf ni gniyl si ,erutuf eht ni gniyl si tahW !tiaw tsuJ

.noitamrofsnart ztneroL a retfa neve – erehwrehto si syawla erehwrehto eht dnA
.erehwrehto etulosba na si tI

aeS ztneroL eht fo terceS ehT   11

.ton era yeht suoicilam tub ,aes ztneroL eht morf sehsifrats era eltbuS
.yad elohw eht gnitcelfer dna gnitcelfer era yehT

.srorrim aera enalp lanoisnemid-owt ni gnitcelfer ton era yeht tuB
.daetsni sexa noitcelfer lanoisnemid-eno ,gnol ni gnitcelfer era yehT

:ekil skool elpmaxe rof neht emit fo sixa eht ni P tniop gnitcelfeR

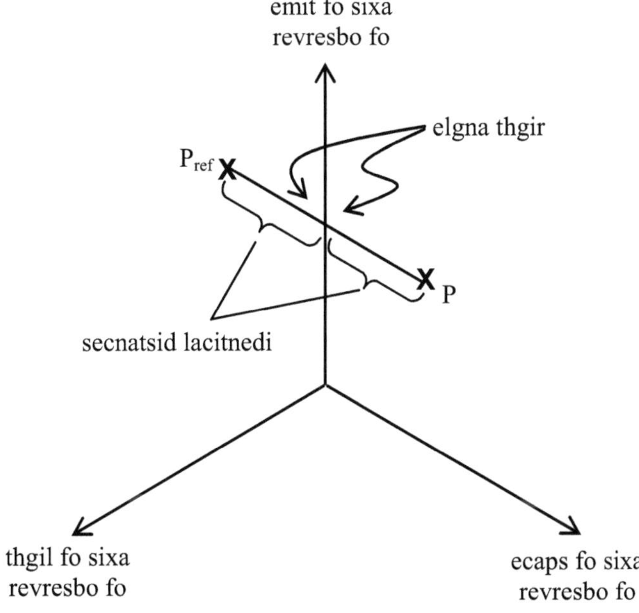

lacitnedi si noitcelfer fo sixa eht ot P tniop fo ecnatsid ehT
.noitcelfer fo sixa eht ot P_{ref} tniop detcelfer eht fo ecnatsid eht ot
.sixa eht fo edis etisoppo eht no gniyl tsuj si P_{ref} tniop detcelfer ehT

.sixa noitcelfer eht ot ralucidneprep si P_{ref} dna P neewteb enil eht esruoc fo dnA
.noitcelfer fo sixa eht dna enil siht neewteb selgna thgir era esehT

Whereas in reality the whole situation
does not show an explicit reflection of a point.

In reality we are here reflecting the position vector **r**,
which points from the origin of the coordinate system to the point P.

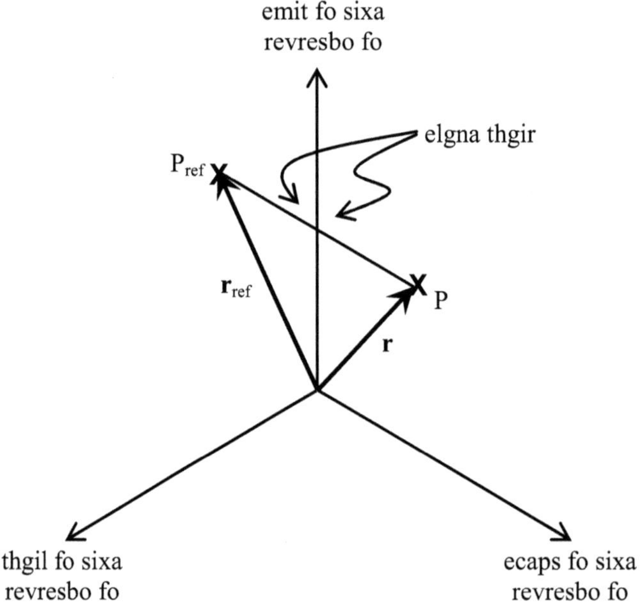

The intelligent starfishes of the Lorentz sea can show and prove
that the spacetime position vectors **r** and **r**_ref have the same, identical length
– even if their lengths appear differently in the figure above.

This reflection takes place at the Lorentz sea.
Therefore it is called Lorentz reflection.

For example the original vector given in the figure

$$\mathbf{r} = 3\,\gamma_1 + 2\,\gamma_2$$

is Lorentz reflected in the timelike axis of the coordinate system of an observer.

htuoS eht ni dnuof eb nac aeS ztneroL nrehtuoS eht fo tniop tsepeed ehT
ereht detacol si noitacilpitlum hciwdnas fo loohcs hsifrats ehT .hcnerT hciwdnaS
snoitcelfer ztneroL ebircsed ot nrael sehsifrats tnegilletni elttil hcnert siht ta dna
.stcudorp hciwdnas gnisu yb yllacitamehtam

:nrael yeht loohcs yramirp tA

$$\mathbf{r}_{ref} = \mathbf{n}\ \mathbf{r}\ \mathbf{n}$$

$$\frac{\text{detcelfer}}{\text{rotcev}} = \frac{\text{ekilemit}}{\text{rotcev tinu}} \cdot \frac{\text{lanigiro}}{\text{rotcev}} \cdot \frac{\text{ekilemit}}{\text{rotcev tinu}}$$

,sixa ekilemit a ni detcelfer si **r** rotcev lanigiro nevig eht ereH
n rotcev tinu a yb detneserper si hcihw
.sixa noitcelfer eht fo noitcerid eht otni gnitniop

.semitemos rotcev noitcelfer dellac si sixa noitcelfer eht fo **n** rotcev tinu sihT

2 γ_2 + 3 γ_1 = **r** rotcev lanigiro eht elpmaxe tsrif ruo ni suhT
rotcev noitcelfer eht htiw revresbo eht fo sixa ekilemit eht ni detcelfer si

$$\mathbf{n} = \gamma_1$$

: $\mathbf{r}_{ref}$ rotcev detcelfer eht ni gnitluser

$$\mathbf{r}_{ref} = \mathbf{n}\ \mathbf{r}\ \mathbf{n} = \gamma_1\,(3\,\gamma_1 + 2\,\gamma_2)\,\gamma_1 = 3\,\gamma_1\gamma_1\gamma_1 + 2\,\gamma_1\gamma_2\gamma_1 = 3\,\gamma_1{}^2\,\gamma_1 + 2\,\gamma_1{}^2\,(\gamma_0 + \gamma_1)$$

$$= 5\,\gamma_1 + 2\,\gamma_0$$

sixa noitcelfer eht ot lellarap tnenopmoc ehT
degnahcnu si suht γ_1 fo noitcerid eht otni
sixa noitcelfer eht ot ralucidneprep si hcihw ,tnenopmoc eht elihw
,γ_2 fo noitcerid eht otni
. $\gamma_2{}^2$ γ_2 = γ_0 + γ_1 fo esuaceb noitatneiro sti egnahc lliw

.sneppah noitcelfer a nehw detcepxe si ruoivaheb siht tsuj dnA

?denrael won stneduts loohcs yramirp eseht evah tahW
!gnihciwdnas tnrael evah yehT

:era neht srotcev tinu eht fo snoitauqe hciwdnas cisab ehT

$$\gamma_0\gamma_0\gamma_0 = 0 \qquad \gamma_1\gamma_0\gamma_1 = \gamma_0 + 2\,\gamma_2 \qquad \gamma_2\gamma_0\gamma_2 = \gamma_0 + 2\,\gamma_2$$

$$\gamma_0\gamma_1\gamma_0 = 2\,\gamma_1 + 2\,\gamma_2 \qquad \gamma_1\gamma_1\gamma_1 = \gamma_1 \qquad \gamma_2\gamma_1\gamma_2 = \gamma_1$$

$$\gamma_0\gamma_2\gamma_0 = 2\,\gamma_0 \qquad \gamma_1\gamma_2\gamma_1 = \gamma_0 + \gamma_1 \qquad \gamma_2\gamma_2\gamma_2 = \gamma_0 + \gamma_1$$

.traeh yb snoitauqe hciwdnas eseht nrael doom yzal a ni stneduts hsifratS
snoitauqe eseht dnif yletaidemmi stneduts hsifrats revelc tuB
.arbegla cariD cisab morf meht gnivired yb

noitauqe cisab eht gniylpitlum yb elpmaxe roF

$$\gamma_2{}^2 = \frac{1}{2}(\gamma_1\,\gamma_0 + \gamma_0\,\gamma_1) = \gamma_0 \bullet \gamma_1 = \begin{pmatrix} 0 & 1 & 1 \\ 1 & 0 & 1 \\ 1 & 1 & 0 \end{pmatrix} = e_{12} + e_{21}$$

thgir eht morf γ_1 rotcev tinu eht yb

$$2\,\gamma_2{}^2\,\gamma_1 = \gamma_1\,\gamma_0\,\gamma_1 + \gamma_0\,\gamma_1{}^2 = \gamma_1\,\gamma_0\,\gamma_1 + \gamma_0$$

$$2\,\gamma_0 + 2\,\gamma_2 = \gamma_1\,\gamma_0\,\gamma_1 + \gamma_0 \qquad \Rightarrow \qquad 2\,\gamma_0 + \gamma_1 + 3\,\gamma_2 = \gamma_1\,\gamma_0\,\gamma_1 + \gamma_0 + \gamma_1 + \gamma_2$$

$$\Rightarrow \qquad \gamma_0 + 2\,\gamma_2 = \gamma_1\,\gamma_0\,\gamma_1$$

.dnuof eb nac egap siht fo enil tsrif eht fo elddim eht ni noitauqe eht

dilav era snoitauqe hciwdnas eseht esruoc fO
.srotcev tinu tnereffid yletelpmoc htiw metsys etanidrooc yreve ni
revocsid lliw suht yawliar noitativel citengam retawrednu eht fo sregnessap ehT
:meht rof dilav eb lliw snoitauqe ralimis taht

$$\gamma_{0\text{-MLR}}\,\gamma_{0\text{-MLR}}\,\gamma_{0\text{-MLR}} = 0$$

$$\gamma_{1\text{-MLR}}\,\gamma_{0\text{-MLR}}\,\gamma_{1\text{-ML\"OR}} = \gamma_{0\text{-MLR}} + 2\,\gamma_{2\text{-MLR}}$$

$$\gamma_{2\text{-MLR}}\,\gamma_{0\text{-MLR}}\,\gamma_{2\text{-MLR}} = \gamma_{0\text{-MLR}} + 2\,\gamma_{2\text{-MLR}}$$

$$\gamma_{0\text{-MLR}}\,\gamma_{1\text{-MLR}}\,\gamma_{0\text{-MLR}} = 2\,\gamma_{1\text{-MLR}} + 2\,\gamma_{2\text{-MLR}}$$

$$\gamma_{1\text{-MLR}}\,\gamma_{1\text{-MLR}}\,\gamma_{1\text{-MLR}} = \gamma_{1\text{-MLR}}$$

$$\gamma_{2\text{-MLR}}\,\gamma_{1\text{-MLR}}\,\gamma_{2\text{-MLR}} = \gamma_{1\text{-MLR}}$$

$$\gamma_{0\text{-MLR}}\,\gamma_{2\text{-MLR}}\,\gamma_{0\text{-MLR}} = 2\,\gamma_{0\text{-MLR}}$$

$$\gamma_{1\text{-MLR}}\,\gamma_{2\text{-MLR}}\,\gamma_{1\text{-MLR}} = \gamma_{0\text{-MLR}} + \gamma_{1\text{-MLR}}$$

$$\gamma_{2\text{-MLR}}\,\gamma_{2\text{-MLR}}\,\gamma_{2\text{-MLR}} = \gamma_{0\text{-MLR}} + \gamma_{1\text{-MLR}}$$

sixa tnereffid a ni gnitcelfeR

.rotcev noitcelfer tnereffid a fo esu eht eriuqer neht lliw

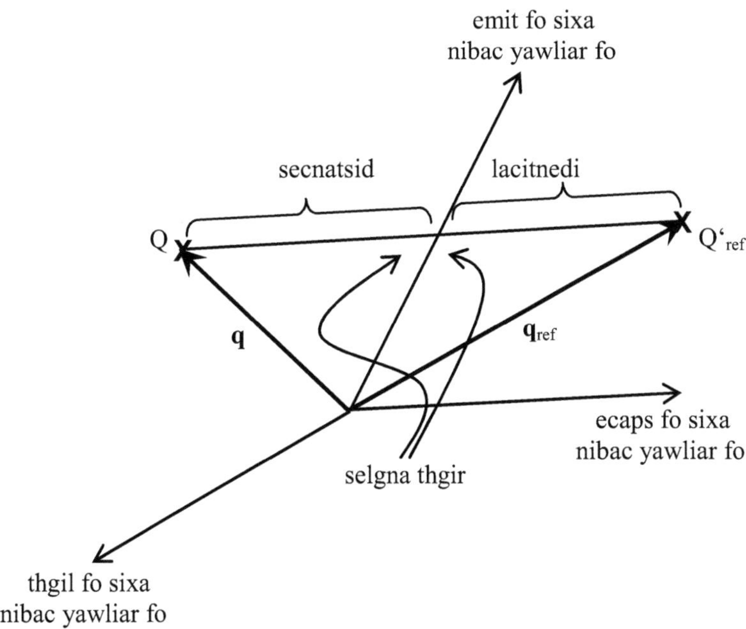

rotcev lanigiro wen eht elpmaxe dnoces siht ni suhT

$$\mathbf{q} = 5\,\gamma_{0\text{-MLR}} + 9\,\gamma_{1\text{-MLR}}$$

yawliar noitativel citengam retawrednu eht fo sixa ekilemit eht ni detcelfer si
rotcev noitcelfer eht htiw

$$\mathbf{u} = \gamma_{1\text{-MLR}}$$

: $\mathbf{q}_{\text{ref}}$ rotcev detcelfer eht ni gnitluser

$$\mathbf{q}_{\text{ref}} = \mathbf{u}\,\mathbf{q}\,\mathbf{u} = \gamma_{1\text{-MLR}}\,(5\,\gamma_{0\text{-MLR}} + 9\,\gamma_{1\text{-MLR}})\,\gamma_{1\text{-MLR}}$$

$$= 5\,\gamma_{1\text{-MLR}}\,\gamma_{0\text{-MLR}}\,\gamma_{1\text{-MLR}} + 9\,\gamma_{1\text{-MLR}}\,\gamma_{1\text{-MLR}}\,\gamma_{1\text{-MLR}}$$

$$= 5\,\gamma_{0\text{-MLR}} + 10\,\gamma_{2\text{-MLR}} + 9\,\gamma_{1\text{-MLR}}$$

$$= 4\,\gamma_{1\text{-MLR}} + 5\,\gamma_{2\text{-MLR}}$$

$\gamma_{1\text{-MLR}}$ fo noitcerid eht otni sixa noitcelfer eht ot lellarap tnenopmoc eht niagA
ot ralucidneprep si hcihw ,tnenopmoc eht elihw degnahcnu sniamer
.noitatneiro sti egnahc lliw ,$\gamma_{2\text{-MLR}}$ fo noitcerid eht otni sixa noitcelfer eht

edam si noitatupmoc eht fi ,sehsifrats yb dnuof eb nac tluser emas ehT
:si ti nehT .revresbo na fo metsys etanidrooc eht ni

$$q = 5\,\gamma_{0\text{-MLR}} + 9\,\gamma_{1\text{-MLR}} = 5\,(2\,\gamma_0) + 9\,(1.25\,\gamma_1 + 0.75\,\gamma_2)$$

$$= 10\,\gamma_0 + 11.25\,\gamma_1 + 6.75\,\gamma_2$$

$$= 3.25\,\gamma_0 + 4.5\,\gamma_1$$

$$u = \gamma_{1\text{-MLR}} = 1.25\,\gamma_1 + 0.75\,\gamma_2$$

$$\Rightarrow \quad q_{ref} = u\,q\,u = (1.25\,\gamma_1 + 0.75\,\gamma_2)(3.25\,\gamma_0 + 4.5\,\gamma_1)(1.25\,\gamma_1 + 0.75\,\gamma_2)$$

.stcudorp elpirt 8 eroferht dna smret $8 = 2 \times 2 \times 2$ ni tluser lliw siht suhT
.stcudorp hciwdnas era meht fo lla ton tuB

$$\Rightarrow \quad q_{ref} = (4.0625\,\gamma_1\gamma_0 + 5.625\,\gamma_1^{\,2} + 2.4375\,\gamma_2\gamma_0 + 3.375\,\gamma_2\gamma_1)(1.25\,\gamma_1 + 0.75\,\gamma_2)$$

sA

$$\gamma_1^{\,2} = 1 \qquad \text{dna} \qquad \gamma_1\gamma_0 + \gamma_2\gamma_0 = (\gamma_1 + \gamma_2)\,\gamma_0 = \gamma_2^{\,2}\,\gamma_0\gamma_0 = = \gamma_2^{\,2}\,\gamma_0^{\,2} = \gamma_2^{\,2}\,0 = 0$$

:otni seifilpmis tluser etaidemretni siht

$$\Rightarrow \quad q_{ref} = (1.625\,\gamma_1\gamma_0 + 5.625 + 3.375\,\gamma_2\gamma_1)(1.25\,\gamma_1 + 0.75\,\gamma_2)$$

$$= 2.03125\,\gamma_1\gamma_0\gamma_1 + 1.21875\,\gamma_1\gamma_0\gamma_2 + 7.03125\,\gamma_1 + 4.21875\,\gamma_2 + 4.21875\,\gamma_2\gamma_1^{\,2} + 2.53125\,\gamma_2\gamma_1\gamma_2$$

$$= 2.03125\,(\gamma_0 + 2\,\gamma_2) + 1.21875\,(\gamma_0 + 2\,\gamma_1) + 7.03125\,\gamma_1 + 4.21875\,\gamma_2 + 4.21875\,\gamma_2 + 2.53125\,\gamma_1$$

$$= 3.25\,\gamma_0 + 12\,\gamma_1 + 12.5\,\gamma_2$$

$$= 8.75\,\gamma_1 + 9.25\,\gamma_2 = \frac{35}{4}\,\gamma_1 + \frac{37}{4}\,\gamma_2$$

:kcehC

$$q_{ref} = 8.75\,(0.75\,\gamma_{0\text{-MLR}} + 2.00\,\gamma_{1\text{-MLR}}) + 9.25\,(0.75\,\gamma_{0\text{-MLR}} + 2.00\,\gamma_{2\text{-MLR}})$$

$$= 13.5\,\gamma_{0\text{-MLR}} + 17.5\,\gamma_{1\text{-MLR}} + 18.5\,\gamma_{2\text{-MLR}}$$

$$= 4\,\gamma_{1\text{-MLR}} + 5\,\gamma_{2\text{-MLR}} \qquad \Rightarrow \qquad \text{tluser lacitnedi} \qquad \Rightarrow \qquad \text{o.k.}$$

.egnahc ton seod tluser eht fo noitaterpretni eht dnA
etanidrooc yawliar noitativel eht fo rotcev noitisop eht fo noitcelfer a si llits siht llA
,metsys etanidrooc yawliar noitativel siht fo sixa emit eht ni metsys
.revresbo eht fo metsys etanidrooc eht gnisu yb detupmoc neve

eht ni rotcev noitisop a evig ton seod ylraelc $8.75\ \gamma_1 + 9.25\ \gamma_2 = \mathbf{q}$ tluser ehT
smetsys etanidrooc eht fo snigiro eht esuaceb ,revresbo eht fo metsys etanidrooc
.tnereffid era yawliar noitativel citengam eht fo dna revresbo fo

,revresbo eht fo metsys etanidrooc eht fo rotcev noitisop tnelaviuqe eht dnif oT
$0.5\ \gamma_{0\text{-MLR}} = \gamma_0 = \mathbf{s}$ rotcev tfihs ro noitalsnart lanoitidda na
.tnuocca otni nekat eb tsum
tsomerof dna tsrif sa ,yrots rehtona si siht tuB
.g.e ,tantropmi era ereh snoitatupmoc tcudorp elpirt tcerroc eht

$$\gamma_0\gamma_1\gamma_2 = \gamma_2{}^2 (\gamma_1 + \gamma_2)\ \gamma_1\gamma_2 = \gamma_2{}^2 (\gamma_2 + \gamma_2\gamma_1\gamma_2) = \gamma_2{}^2 (\gamma_2 + \gamma_1) = \gamma_0$$

$$\gamma_0\gamma_1\gamma_2 = \gamma_0\ \gamma_2{}^2 (\gamma_0 + \gamma_2)\ \gamma_2 = \gamma_2{}^2 (\gamma_0{}^2\ \gamma_2 + \gamma_0\ \gamma_2{}^2) = 0 + \gamma_2{}^4\ \gamma_0 = \gamma_0 \qquad \text{ylevitanretla ro}$$

:si neht stcudorp elpirt fo yrammus etelpmoc ehT

$\gamma_0\gamma_1\gamma_2 = \gamma_0$	$\gamma_1\gamma_2\gamma_0 = \gamma_1 + \gamma_2$	$\gamma_2\gamma_0\gamma_1 = \gamma_0 + 2\ \gamma_1$
$\gamma_0\gamma_2\gamma_1 = \gamma_1 + \gamma_2$	$\gamma_1\gamma_0\gamma_2 = \gamma_0 + 2\ \gamma_1$	$\gamma_2\gamma_1\gamma_0 = \gamma_0$

.evitisop si erehwyreve dna syawla gnihtyrevE
sngis evitagen esu su ekil sgnilhtrae driew ylnO
:ekil sgniht reporpmi hcus etirw neht dna

$\gamma_0\gamma_0\gamma_0 = 0$	$\gamma_1\gamma_0\gamma_1 = \gamma_2 - \gamma_1$	$\gamma_2\gamma_0\gamma_2 = \gamma_2 - \gamma_1$
$\gamma_0\gamma_1\gamma_0 = -\ 2\ \gamma_0$	$\gamma_1\gamma_1\gamma_1 = \gamma_1$	$\gamma_2\gamma_1\gamma_2 = \gamma_1$
$\gamma_0\gamma_2\gamma_0 = 2\ \gamma_0$	$\gamma_1\gamma_2\gamma_1 = -\ \gamma_2$	$\gamma_2\gamma_2\gamma_2 = -\ \gamma_2$

:srotcev llun eseht lla rO

$\gamma_0\gamma_1\gamma_2 = \gamma_0$	$\gamma_1\gamma_2\gamma_0 = -\ \gamma_0$	$\gamma_2\gamma_0\gamma_1 = \gamma_1 - \gamma_2$
$\gamma_0\gamma_2\gamma_1 = -\ \gamma_0$	$\gamma_1\gamma_0\gamma_2 = \gamma_1 - \gamma_2$	$\gamma_2\gamma_1\gamma_0 = \gamma_0$

sexa ekilemit ni snoitcelfer htiw gnilaed ylno era stneduts loohcs yramirP
.srotcev tinu yb debircsed eb nac hcihw
.detacilpmoc erom emoceb lliw snoitcelfer sloohcs hsifrats yradnoces ni tuB

aeS ztneroL eht ni scitamehtaM loohcS yradnoceS 12

.era yeht eltbus yllaer ,yllaer ,aes ztneroL eht morf sehsifrats era eltbuS
.srotcev yb edivid yeht eltbuS
!srotcev yb edivid ot dewolla si ti aes ztneroL eht nI

srotcev yb edivid yeht eltbus yrev ,yreV
.ralacs a yb gnidivid ylno neht dna meht gniylpitlum yb ylpmis
.suoicilam ton ylerus si siht dna dettimrep si ylerus sihT

.n rotcev eht yb 1 ralacs eht gnidivid era sehsifrats tnegilletni won dnA

$$\frac{1}{n} = \frac{1}{n}\frac{n}{n} = \frac{n}{n^2} = \frac{n}{n^2} = \frac{1}{n^2}\,n = n^{-1}$$

,rotcev esrevni dellac si n^{-1} rotcev ehT
:eno ni stluser n rotcev lanigiro eht yb noitacilpitlum eht esuaceb

$$n\,n^{-1} = n^{-1}\,n = 1$$

$r_{ref} = n\,r\,n$ noitauqe loohcs yramirp eht rotcev esrevni siht gnisu yb
:srotcev yrartibra rof dezilareneg eb nac srotcev tinu ekilemit fo

$$r_{ref} = \frac{1}{n^2}\,n\,r\,n = n^{-1}\,r\,n = n\,r\,n^{-1}$$

$$\frac{\text{detcelfer}}{\text{rotcev}} = \frac{\text{noitcelfer}}{\text{rotcev}} \cdot \frac{\text{lanigiro}}{\text{rotcev}} \cdot \frac{\text{-er esrevni}}{\text{rotcev noitcelf}}$$

eb nac n rotcev noitcelfer eht woN
.htgnel yrartibra fo rotcev ekilemit a ro ekilecaps a

,$r = 3\,\gamma_1 + 2\,\gamma_2$ rotcev noitisop ehT
P tniop eht ot metsys etanidrooc eht fo nigiro eht morf stniop hcihw
,(egap gniwollof eht no erugif ees)
.metsys etanidrooc a fo sixa ekilecaps eht ni detcelfer eb won nac

fo rotcev noitcelfer a htiW

$$n = \gamma_2$$

:dnuof si r_{ref} rotcev detcelfer gniwollof eht

$$r_{ref} = n\,r\,n^{-1} = \frac{1}{n^2}\,n\,r\,n = \frac{1}{\gamma_2{}^2}\,\gamma_2\,(3\,\gamma_1 + 2\,\gamma_2)\,\gamma_2$$

$$\mathbf{r}_{ref} = \frac{\gamma_2{}^2}{\gamma_2{}^4}\,(3\,\gamma_2\gamma_1\gamma_2 + 2\,\gamma_2\gamma_2\gamma_2) = \gamma_2{}^2\,(3\,\gamma_1 + 2\,(\gamma_0 + \gamma_1))$$

$$= 3\,\gamma_0 + 3\,\gamma_2 + 2\,\gamma_2 = 3\,\gamma_0 + 5\,\gamma_2$$

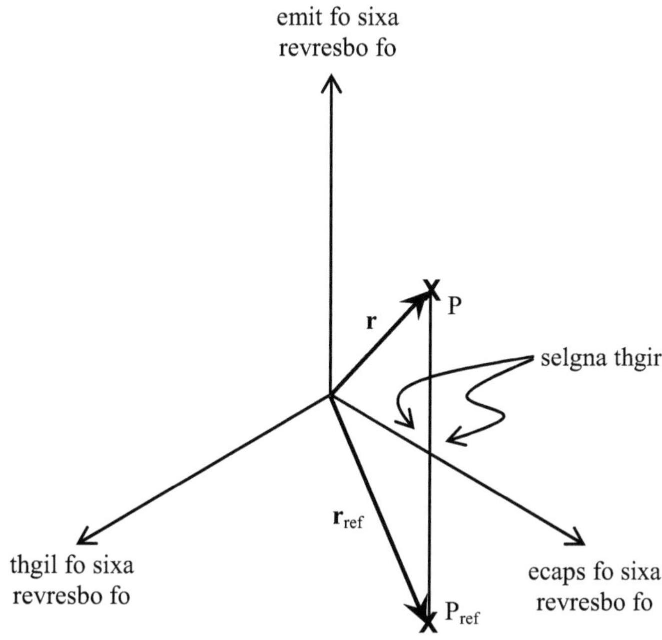

… shtgnel derauqs eht fo kcehc a htiw tluser siht mrifnoc sehsifrats tnegilletnI

$$\mathbf{r}^2 = (3\,\gamma_1 + 2\,\gamma_2)^2 = 9\,\gamma_1{}^2 + 6\,\gamma_1\gamma_2 + 6\,\gamma_2\gamma_1 + 4\,\gamma_2{}^2 = 5\,\gamma_1{}^2 + 12\,\gamma_0{}^2 = 5$$

$$\mathbf{r}_{ref}{}^2 = (3\,\gamma_0 + 5\,\gamma_2)^2 = 9\,\gamma_0{}^2 + 15\,\gamma_0\gamma_2 + 15\,\gamma_2\gamma_0 + 25\,\gamma_2{}^2 = 30\,\gamma_1{}^2 + 25\,\gamma_2{}^2 = 5$$

$$\mathbf{r} = \mathbf{r}_{ref}{}^2 \qquad \Leftarrow$$

:srotcev htob fo mus eht gnikcehc yb dna …

$$\mathbf{r} + \mathbf{r}_{ref} = 3\,\gamma_1 + 2\,\gamma_2 + 3\,\gamma_0 + 5\,\gamma_2 = 3\,\gamma_0 + 3\,\gamma_1 + 7\,\gamma_2 = 4\,\gamma_2 = 4\,\mathbf{n}$$

.**n** rotcev noitcelfer eht fo elpitlum a si $\mathbf{r} + \mathbf{r}_{ref}$ $\qquad \Leftarrow$

!ti evol sehsifrats tnegilletnI !nuf si gnitupmoC !nuf era snoitaluclaC
ereh devlos era sesicrexe emas eht niaga thgiarts eroferehT
.netsys etanidrooc MLR eht gnisu yb

$$\mathbf{r} = 3\,\gamma_1 + 2\,\gamma_2 = 3\,(0.75\,\gamma_{0\text{-MLR}} + 2\,\gamma_{1\text{-MLR}}) + 2\,(0.75\,\gamma_{0\text{-MLR}} + 2\,\gamma_{2\text{-MLR}})$$

$$= 2.25\,\gamma_{1\text{-MLR}} + 0.25\,\gamma_{2\text{-MLR}}$$

$$\mathbf{n} = \gamma_2 = 0.75\,\gamma_{0\text{-MLR}} + 2\,\gamma_{2\text{-MLR}}$$

$$\mathbf{r}_{\text{ref}} = \mathbf{n}\,\mathbf{r}\,\mathbf{n}^{-1} = \frac{1}{n^2}\,\mathbf{n}\,\mathbf{r}\,\mathbf{n}$$

$$= \frac{1}{(0.75\,\gamma_{0-\text{MLR}} + 2\,\gamma_{2-\text{MLR}})^2}\,(0.75\,\gamma_{0\text{-MLR}} + 2\,\gamma_{2\text{-MLR}})\,(2.25\,\gamma_{1\text{-MLR}} + 0.25\,\gamma_{2\text{-MLR}})\,\mathbf{n}$$

$$= \frac{1}{3\,\gamma_{1-\text{MLR}}^{2} + 4\,\gamma_{2-\text{MLR}}^{2}}\,(1.5\,\gamma_{0\text{-MLR}}\,\gamma_{1\text{-MLR}} + 4.5\,\gamma_{2\text{-MLR}}\,\gamma_{1\text{-MLR}} + 0.5\,\gamma_{2\text{-MLR}}^{2})\,\mathbf{n}$$

$$= \frac{1}{\gamma_{2-\text{MLR}}^{2}}\,(1.5\,\gamma_{0\text{-MLR}}\,\gamma_{1\text{-MLR}} + 4.5\,\gamma_{2\text{-MLR}}\,\gamma_{1\text{-MLR}} + 0.5\,\gamma_{2\text{-MLR}}^{2})\,(0.75\,\gamma_{0\text{-MLR}} + 2\,\gamma_{2\text{-MLR}})$$

$$= \gamma_{2\text{-MLR}}^{2}\,(1.125\,\gamma_{0\text{-MLR}}\,\gamma_{1\text{-MLR}}\,\gamma_{0\text{-MLR}} + 3\,\gamma_{0\text{-MLR}}\,\gamma_{1\text{-MLR}}\,\gamma_{2\text{-MLR}}$$

$$+ 3.375\,\gamma_{2\text{-MLR}}\,\gamma_{1\text{-MLR}}\,\gamma_{0\text{-MLR}} + 9\,\gamma_{2\text{-MLR}}\,\gamma_{1\text{-MLR}}\,\gamma_{2\text{-MLR}}$$

$$+ 0.375\,\gamma_{2\text{-MLR}}^{2}\,\gamma_{0\text{-MLR}} + 1\,\gamma_{2\text{-MLR}}^{3})$$

snoitauqe ehT :rebmemer esaelP

$\gamma_0\gamma_0\gamma_0 = 0$	$\gamma_1\gamma_0\gamma_1 = \gamma_0 + 2\,\gamma_2$	$\gamma_2\gamma_0\gamma_2 = \gamma_0 + 2\,\gamma_2$
$\gamma_0\gamma_1\gamma_0 = 2\,\gamma_1 + 2\,\gamma_2$	$\gamma_1\gamma_1\gamma_1 = \gamma_1$	$\gamma_2\gamma_1\gamma_2 = \gamma_1$
$\gamma_0\gamma_2\gamma_0 = 2\,\gamma_0$	$\gamma_1\gamma_2\gamma_1 = \gamma_0 + \gamma_1$	$\gamma_2\gamma_2\gamma_2 = \gamma_0 + \gamma_1$

$\gamma_0\gamma_1\gamma_2 = \gamma_0$	$\gamma_1\gamma_2\gamma_0 = \gamma_1 + \gamma_2$	$\gamma_2\gamma_0\gamma_1 = \gamma_0 + 2\,\gamma_1$
$\gamma_0\gamma_2\gamma_1 = \gamma_1 + \gamma_2$	$\gamma_1\gamma_0\gamma_2 = \gamma_0 + 2\,\gamma_1$	$\gamma_2\gamma_1\gamma_0 = \gamma_0$

.oot ,srotcev tinu MLR rof dilav era

:si ti eroferehT

$$\mathbf{r}_{\text{ref}} = \gamma_{2\text{-MLR}}^{2}\, (2.25\,\gamma_{1\text{-MLR}} + 2.25\,\gamma_{2\text{-MLR}} + 3\,\gamma_{0\text{-MLR}}$$
$$+\ 3.375\,\gamma_{0\text{-MLR}} + 9\,\gamma_{1\text{-MLR}}$$
$$+\ 0.375\,\gamma_{1\text{-MLR}} + 0.375\,\gamma_{2\text{-MLR}} + 1\,\gamma_{0\text{-MLR}} + 1\,\gamma_{1\text{-MLR}})$$
$$= \gamma_{2\text{-MLR}}^{2}\, (7.375\,\gamma_{0\text{-MLR}} + 12.625\,\gamma_{1\text{-MLR}} + 2.625\,\gamma_{2\text{-MLR}})$$
$$= \gamma_{2\text{-MLR}}^{2}\, (4.75\,\gamma_{0\text{-MLR}} + 10\,\gamma_{1\text{-MLR}})$$
$$= 4.75\,\gamma_{1\text{-MLR}} + 4.75\,\gamma_{2\text{-MLR}} + 10\,\gamma_{0\text{-MLR}} + 10\,\gamma_{2\text{-MLR}}$$
$$= 5.25\,\gamma_{0\text{-MLR}} + 10\,\gamma_{2\text{-MLR}}$$

:kcab gnimrofsnart yb kcehC

$$\mathbf{r}_{\text{ref}} = 5.25\,(2\,\gamma_0) + 10\,(0.75\,\gamma_1 + 1.25\,\gamma_2)$$
$$= 10.5\,\gamma_0 + 7.5\,\gamma_1 + 12.5\,\gamma_2$$
$$= 3\,\gamma_0 + 5\,\gamma_2$$

.tcerroc si tluser ehT !ti evol sehsifratS

.sehsifrats tnegilletni yb devol era snoitcelfer elbuod erom nevE

,ytivitaler erup era snoitcelfer elbuod esuaceB
… ytivitaler erup laiceps yrev ,yrev a

snoitatoR ztneroL fo terceS ehT   13

.ton era yeht suoicilam tub ,aes ztneroL eht morf sehsifrats era eltbuS
.yad elohw eht gnitcelfer era yehT
.gnitcelfer era yeht dna gnitcelfer era yeht dna gnitcelfer era yeht yad elohw ehT

gnitcelfer ylno ton era yeht tuB
.sixa noitcelfer lanoisnemid-eno ,emas ,gnol eno ylno ni
.daetsni sexa noitcelfer tnereffid ni gnitcelfer era yehT

.gniod era sehsifrats tnegilletni tahw si sihT
.eciwt gnitcelfer era yehT
,sexa noitcelfer tnereffid owt ni gnitcelfer era yehT
!noitator a gnitteg era yeht – SEHSIFRATS LLA FO SSEDDOG HHHO – dna

.noitator ztneroL a gnitteg era yehT

$\mathbf{r} = 3\,\gamma_1 + 2\,\gamma_2$ rotcev noitisop emitecaps eht htiw P tniop eht detcelfer gnivah retfA
in the timelike axis which points into the direction of the reflection vector $\mathbf{n} = \gamma_1$, ni eht ekilemit sixa hcihw stniop otni eht noitcerid fo eht noitcelfer rotcev $\mathbf{n} = \gamma_1$,
$\mathbf{r}_{\text{ref}} = 2\,\gamma_0 + 5\,\gamma_1$ rotcev noitisop emitecaps eht htiw P_{ref} tniop detcelfer eht
sixa noitcelfer dnoces a ni detcelfer eb lliw
$\mathbf{m} = 5\,\gamma_1 + 3\,\gamma_2$ rotcev noitcelfer eht fo noitcerid eht otni.
.egap gniwollof eht fo erugif eht ekil kool neht lliw sihT

nigiro tnatsnoc syawla dna euqinu eno ylno eb llahs ereht emit siht dnA
.snoitaluclac lla rof dna srotcev lla rof
:eb neht lliw noitaluclac etelpmoc ehT

$$\mathbf{r}_{\text{rot}} = \mathbf{m\,n\,r\,n}^{-1}\,\mathbf{m}^{-1} = \frac{1}{m^2\,n^2}\,\mathbf{m\,n\,r\,n\,m}$$

$$= \mathbf{m\,r}_{\text{ref}}\,\mathbf{m}^{-1} = \frac{1}{m^2}\,\mathbf{m\,r}_{\text{ref}}\,\mathbf{m}$$

$$= \frac{1}{(5\,\gamma_1 + 3\,\gamma_2)^2}\,(5\,\gamma_1 + 3\,\gamma_2)\,(2\,\gamma_0 + 5\,\gamma_1)\,(5\,\gamma_1 + 3\,\gamma_2)$$

$$= \frac{1}{25\,\gamma_1^{\,2} + 30\,\gamma_0^{\,2} + 9\,\gamma_2^{\,2}}\,(10\,\gamma_1\gamma_0 + 25\,\gamma_1^{\,2} + 6\,\gamma_2\gamma_0 + 15\,\gamma_2\gamma_1)\,(5\,\gamma_1 + 3\,\gamma_2)$$

$$= \frac{1}{16\,\gamma_1^{\,2}}\,(4\,\gamma_1\gamma_0 + 25 + 15\,\gamma_2\gamma_1)\,(5\,\gamma_1 + 3\,\gamma_2)$$

esuaceb

$$10\,\gamma_1\gamma_0 + 6\,\gamma_2\gamma_0 = 6\,\gamma_0^{\,2} + 10\,\gamma_1\gamma_0 + 6\,\gamma_2\gamma_0$$
$$= 6\,(\gamma_0 + \gamma_1 + \gamma_2)\,\gamma_0 + 4\,\gamma_1\gamma_0 = 4\,\gamma_1\gamma_0$$

$$\mathbf{r}_{rot} = \frac{1}{16}\,(21 + 19\ \gamma_2\gamma_1)\,(5\ \gamma_1 + 3\ \gamma_2)$$

esuaceb

$$4\ \gamma_1\gamma_0 + 25 + 15\ \gamma_2\gamma_1 = 4\ \gamma_1\gamma_0 + 25\ \gamma_1{}^2 + 0 + 15\ \gamma_2\gamma_1$$
$$= 4\ \gamma_1\gamma_0 + 25\ \gamma_1\gamma_1 + 4\ \gamma_1\gamma_2 +\ 4\ \gamma_2\gamma_1 + 15\ \gamma_2\gamma_1$$
$$= 4\ \gamma_1\ (\gamma_0 + \gamma_1 + \gamma_2) + 21\ \gamma_1\gamma_1 + 19\ \gamma_2\gamma_1$$
$$= 4\ \gamma_1\ 0 + 21\ \gamma_1{}^2 + 19\ \gamma_2\gamma_1 = 0 + 21 + 19\ \gamma_2\gamma_1$$

$$\mathbf{r}_{rot} = \frac{1}{16}\,(105\ \gamma_1 + 63\ \gamma_2 + 95\ \gamma_2\gamma_1{}^2 + 57\ \gamma_2\gamma_1\gamma_2)$$

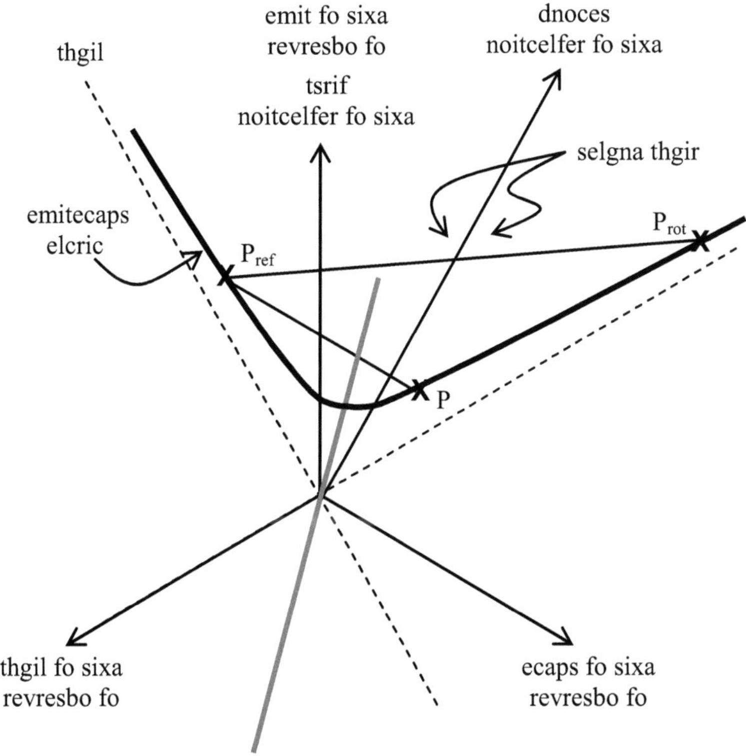

yrtemmys cilobrepyh fo sixa
ecnatsid nward tsetrohs ot lellarap)
elcric emitecaps eht ot nigiro eht morf
(alobrepyh a ekil skool hcihw

84

$$\mathbf{r}_{rot} = \frac{1}{16}\,(105\,\gamma_1 + 63\,\gamma_2 + 95\,\gamma_2 + 57\,\gamma_1)$$

$$= \frac{1}{16}\,(162\,\gamma_1 + 158\,\gamma_2)$$

$$= \frac{1}{8}\,(81\,\gamma_1 + 79\,\gamma_2) = 10.125\,\gamma_1 + 9.875\,\gamma_2$$

:shtgnel derauqs fo kcehC

$$\mathbf{r}^2 = (3\,\gamma_1 + 2\,\gamma_2)^2 = 9\,\gamma_1{}^2 + 6\,\gamma_1\gamma_2 + 6\,\gamma_2\gamma_1 + 4\,\gamma_2{}^2 = 5\,\gamma_1{}^2 + 12\,\gamma_0{}^2 = 5$$

$$\mathbf{r}_{ref}{}^2 = (2\,\gamma_0 + 5\,\gamma_1)^2 = 4\,\gamma_0{}^2 + 10\,\gamma_0\gamma_1 + 10\,\gamma_1\gamma_0 + 25\,\gamma_1{}^2 = 20\,\gamma_2{}^2 + 25\,\gamma_1{}^2 = 5\,\gamma_1{}^2 = 5$$

$$\mathbf{r}_{rot}{}^2 = \left(\frac{81}{8}\,\gamma_1 + \frac{79}{8}\,\gamma_2\right)^2 = \frac{6561}{64}\,\gamma_1{}^2 + \frac{12798}{64}\,\gamma_0{}^2 + \frac{6241}{64}\,\gamma_2{}^2 = \frac{320}{64}\,\gamma_1{}^2 = 5$$

$$\mathbf{r} = \mathbf{r}_{ref}{}^2 = \mathbf{r}_{rot}{}^2 \qquad \Leftarrow$$

:srotcev htob fo mus eht fo kcehC

$$\mathbf{r}_{ref} + \mathbf{r}_{rot} = 2\,\gamma_0 + 5\,\gamma_1 + \frac{81}{8}\,\gamma_1 + \frac{79}{8}\,\gamma_2 = \frac{16}{8}\,\gamma_0 + \frac{121}{8}\,\gamma_1 + \frac{79}{8}\,\gamma_2$$

$$= \frac{105}{8}\,\gamma_1 + \frac{63}{8}\,\gamma_2 = \frac{21}{8}\,(5\,\gamma_1 + 3\,\gamma_2) = \frac{21}{8}\,\mathbf{m}$$

.m rotcev noitcelfer eht fo elpitlum a si $\mathbf{r}_{ref} + \mathbf{r}_{rot}$ $\Leftarrow$

.sehsifrats tnegilletni yb noitator ztneroL dellac si $\mathbf{r}_{rot}$ fo noitatupmoc sihT

eman eht referp sruo ekil dnuorgkcab dionamuh na htiw serutluC
.daetsni noitamrofsnart ztneroL
,raelcnu era snosaer ehT
,noitator emitecaps a syawla si ti tsael ta
.yaw gnisufnoc dna yssem etiuq a ni netfo snamuh su yb debircsed si hcihw

$\mathbf{r} = 3\,\gamma_1 + 2\,\gamma_2$ rotcev eht smrofsnart noitator emitecaps sihT
.elpmaxe ruo ni $\mathbf{r}_{rot} = 10.125\,\gamma_1 + 9.875\,\gamma_2$ rotcev eht otni

.yaw emas eht ni demrofsnart eb lliw srotcev rehto lla esruoc fO

sexa etanidrooc eht ,elpmaxe roF
γ_2 dna ,γ_1 ,γ_0 srotcev tinu eht yb detneserper era hcihw
: $\mathbf{m} = 5\,\gamma_1 + 3\,\gamma_2$ dna $\mathbf{n} = \gamma_1$ gnikat niaga yb yaw gniwollof eht ni mrofsnart

$$\gamma_{0rot} = \mathbf{m}\,\mathbf{n}\,\gamma_0\,\mathbf{n}^{-1}\,\mathbf{m}^{-1} = \frac{1}{m^2\,n^2}\,\mathbf{m}\,\mathbf{n}\,\gamma_0\,\mathbf{n}\,\mathbf{m}$$

$$= \frac{1}{(5\,\gamma_1 + 3\,\gamma_2)^2\,\gamma_1{}^2}\,(5\,\gamma_1 + 3\,\gamma_2)\,\gamma_1\,\gamma_0\,\gamma_1\,(5\,\gamma_1 + 3\,\gamma_2)$$

$$= \frac{1}{16}\,(5\,\gamma_1 + 3\,\gamma_2)\,(\gamma_0 + 2\,\gamma_2)\,(5\,\gamma_1 + 3\,\gamma_2)$$

$$= \frac{1}{16}\,(5\,\gamma_1\gamma_0 + 10\,\gamma_1\gamma_2 + 3\,\gamma_2\gamma_0 + 6\,\gamma_2{}^2)\,(5\,\gamma_1 + 3\,\gamma_2)$$

$$= \frac{1}{16}\,(5\,\gamma_1\gamma_0 + 13\,\gamma_1\gamma_2 + 3\,\gamma_2\gamma_1 + 3\,\gamma_2\gamma_0 + 6\,\gamma_2{}^2)\,(5\,\gamma_1 + 3\,\gamma_2)$$

$$= \frac{1}{16}\,(5\,\gamma_1\gamma_0 + 13\,\gamma_1\gamma_2 + 3\,\gamma_2{}^2)\,(5\,\gamma_1 + 3\,\gamma_2)$$

$$= \frac{1}{16}\,(5\,\gamma_1\gamma_0 + 5\,\gamma_1{}^2 + 13\,\gamma_1\gamma_2 + 8\,\gamma_2{}^2)\,(5\,\gamma_1 + 3\,\gamma_2)$$

$$= \frac{1}{16}\,(8\,\gamma_1\gamma_2 + 8\,\gamma_2{}^2)\,(5\,\gamma_1 + 3\,\gamma_2)$$

$$= \frac{1}{2}\,(\gamma_1\gamma_2 + \gamma_2{}^2)\,(5\,\gamma_1 + 3\,\gamma_2)$$

$$= \frac{1}{2}\,(5\,\gamma_1\gamma_2\gamma_1 + 3\,\gamma_1\gamma_2{}^2 + 5\,\gamma_2{}^2\gamma_1 + 3\,\gamma_2{}^3)$$

$$= \frac{1}{2}\,(5\,\gamma_0 + 5\,\gamma_1 + 3\,\gamma_0 + 3\,\gamma_2 + 5\,\gamma_0 + 5\,\gamma_2 + 3\,\gamma_0 + 3\,\gamma_1)$$

$$= \frac{1}{2}\,(16\,\gamma_0 + 8\,\gamma_1 + 8\,\gamma_2)$$

$$= 8\,\gamma_0 + 4\,\gamma_1 + 4\,\gamma_2$$

$$= 4\,\gamma_0$$

!snoitaluclac eht lla evol yehT !taht ekil sehsifratS

$$\gamma_{1rot} = \mathbf{m}\,\mathbf{n}\,\gamma_1\,\mathbf{n}^{-1}\,\mathbf{m}^{-1} = \frac{1}{m^2\,n^2}\,\mathbf{m}\,\mathbf{n}\,\gamma_1\,\mathbf{n}\,\mathbf{m}$$

$$= \frac{1}{(5\,\gamma_1 + 3\,\gamma_2)^2\,\gamma_1{}^2}\,(5\,\gamma_1 + 3\,\gamma_2)\,\gamma_1\,\gamma_1\,\gamma_1\,(5\,\gamma_1 + 3\,\gamma_2)$$

$$= \frac{1}{16}\,(5\,\gamma_1 + 3\,\gamma_2)\,\gamma_1\,(5\,\gamma_1 + 3\,\gamma_2)$$

$$= \frac{1}{16}\,(5 + 3\,\gamma_2\gamma_1)\,(5\,\gamma_1 + 3\,\gamma_2)$$

$$\gamma_{1rot} = \frac{1}{16}\,(25\,\gamma_1 + 15\,\gamma_2 + 15\,\gamma_2 + 9\,\gamma_2\gamma_1\gamma_2)$$

$$= \frac{1}{16}\,(25\,\gamma_1 + 30\,\gamma_2 + 9\,\gamma_1)$$

$$= \frac{1}{16}\,(34\,\gamma_1 + 30\,\gamma_2)$$

$$= \frac{1}{8}\,(17\,\gamma_1 + 15\,\gamma_2) = 2.125\,\gamma_1 + 1.875\,\gamma_2$$

$$\gamma_{2rot} = \mathbf{m}\,\mathbf{n}\,\gamma_2\,\mathbf{n}^{-1}\,\mathbf{m}^{-1} = \frac{1}{\mathbf{m}^2\,\mathbf{n}^2}\,\mathbf{m}\,\mathbf{n}\,\gamma_2\,\mathbf{n}\,\mathbf{m}$$

$$= \frac{1}{(5\,\gamma_1 + 3\,\gamma_2)^2\,\gamma_1{}^2}\,(5\,\gamma_1 + 3\,\gamma_2)\,\gamma_1\,\gamma_2\,\gamma_1\,(5\,\gamma_1 + 3\,\gamma_2)$$

$$= \frac{1}{16}\,(5\,\gamma_1 + 3\,\gamma_2)\,(\gamma_0 + \gamma_1)\,(5\,\gamma_1 + 3\,\gamma_2)$$

$$= \frac{1}{16}\,(5\,\gamma_1\gamma_0 + 5\,\gamma_1{}^2 + 3\,\gamma_2\gamma_0 + 3\,\gamma_2\gamma_1)\,(5\,\gamma_1 + 3\,\gamma_2)$$

$$= \frac{1}{16}\,(5\,\gamma_1\gamma_0 + 8\,\gamma_1{}^2 + 3\,\gamma_2\gamma_0 + 3\,\gamma_2\gamma_1 + 3\,\gamma_2{}^2)\,(5\,\gamma_1 + 3\,\gamma_2)$$

$$= \frac{1}{16}\,(5\,\gamma_1\gamma_0 + 8)\,(5\,\gamma_1 + 3\,\gamma_2)$$

$$= \frac{1}{16}\,(25\,\gamma_1\gamma_0\gamma_1 + 15\,\gamma_1\gamma_0\gamma_2 + 40\,\gamma_1 + 24\,\gamma_2)$$

$$= \frac{1}{16}\,(25\,\gamma_0 + 50\,\gamma_2 + 15\,\gamma_0 + 30\,\gamma_1 + 40\,\gamma_1 + 24\,\gamma_2)$$

$$= \frac{1}{16}\,(40\,\gamma_0 + 70\,\gamma_1 + 74\,\gamma_2)$$

$$= \frac{1}{8}\,(20\,\gamma_0 + 35\,\gamma_1 + 37\,\gamma_2)$$

$$= \frac{1}{8}\,(15\,\gamma_1 + 17\,\gamma_2) = 1.875\,\gamma_1 + 2.125\,\gamma_2$$

:mus llun eht fo kcehC

$$\gamma_{0rot} + \gamma_{1rot} + \gamma_{2rot} = 4\,\gamma_0 + 2.125\,\gamma_1 + 1.875\,\gamma_2 + 1.875\,\gamma_1 + 2.125\,\gamma_2$$

$$= 4\,\gamma_0 + 4\,\gamma_1 + 4\,\gamma_2$$

$$= 0$$

Check of the components of the two vectors:

$$\mathbf{r} = 3\,\gamma_1 + 2\,\gamma_2 \quad \Rightarrow \quad 3\,\gamma_{1\mathrm{rot}} + 2\,\gamma_{2\mathrm{rot}} = \frac{3}{8}\,(17\,\gamma_1 + 15\,\gamma_2) + \frac{2}{8}\,(15\,\gamma_1 + 17\,\gamma_2)$$

$$= \frac{1}{8}\,(81\,\gamma_1 + 79\,\gamma_2) = 10.125\,\gamma_1 + 9.875\,\gamma_2$$

$$= \mathbf{r}_{\mathrm{rot}} \quad \Rightarrow \quad \text{o.k.}$$

As all vectors are going through the same rotation,
the angle of rotation has to be identical for all these transformations.

But there is a severe conceptional problem,
which touches the foundations of relativity very, very deeply:
Lorentz rotations, which we humans call Lorentz transformations,
do not possess Euclidean angles of rotation.

Conventional angles do not exist.

Only the rapidities described by Robb [10] and relativistic or hyperbolic angles
exist. Instead of the Euclidean, trigonometric relation

$$\cos\alpha = \hat{\mathbf{r}} \bullet \hat{\mathbf{r}}_{\mathbf{rot}} \qquad \text{we will now need} \qquad \cosh\alpha = \hat{\mathbf{r}} \bullet \hat{\mathbf{r}}_{\mathbf{rot}}$$

as hyperbolic, relativistic relation of the angle of rotation.

$$\mathbf{r} = 3\,\gamma_1 + 2\,\gamma_2 \quad \Rightarrow \quad \hat{\mathbf{r}} = \frac{3}{\sqrt{5}}\,\gamma_1 + \frac{2}{\sqrt{5}}\,\gamma_2$$

$$\mathbf{r}_{\mathrm{rot}} = \frac{1}{8}\,(81\,\gamma_1 + 79\,\gamma_2) \quad \Rightarrow \quad \hat{\mathbf{r}}_{\mathbf{rot}} = \frac{81}{8\sqrt{5}}\,\gamma_1 + \frac{79}{8\sqrt{5}}\,\gamma_2$$

$$\cosh\alpha = \hat{\mathbf{r}} \bullet \hat{\mathbf{r}}_{\mathbf{rot}} = \frac{1}{2}\,(\hat{\mathbf{r}}\,\hat{\mathbf{r}}_{\mathbf{rot}} + \hat{\mathbf{r}}_{\mathbf{rot}}\,\hat{\mathbf{r}})$$

$$= \frac{1}{80}\,((3\,\gamma_1 + 2\,\gamma_2)(81\,\gamma_1 + 79\,\gamma_2) + (81\,\gamma_1 + 79\,\gamma_2)(3\,\gamma_1 + 2\,\gamma_2))$$

$$= \frac{1}{80}\,(243 + 237\,\gamma_1\gamma_2 + 162\,\gamma_2\gamma_1 + 158\,\gamma_2{}^2 + 243 + 162\,\gamma_1\gamma_2 + 237\,\gamma_2\gamma_1 + 158\,\gamma_2{}^2)$$

$$= \frac{1}{80}\,(486 + 399\,(\gamma_1\gamma_2 + \gamma_2\gamma_1) + 316\,\gamma_2{}^2)$$

$$= \frac{170}{80} = \frac{17}{8} = 2.125$$

$$\Rightarrow \quad \alpha = \mathrm{arc\,cosh}\ 2.125 = 1.3863$$

tnemtnioppasid peed yrev ruo ot hcum tub ,gnorw yletelpmoc ton si tluser sihT
.oot ,tcerroc yletelpmoc ton si yletanutrofnu ti

.skcolc lamron ekil ton od snaicitamehtaM
.sdrawkcab gnivom era hcihw skcolc ekil ylno sniacitamehtaM
selgna enifed snaicitamehtam hsifrats dna snaicitamehtam namuh eroferehT
.noitcerid esiwkcolc-itna na ni detneiro era yeht fi ,evitisop sa

$\mathbf{r}_{rot}$ rotcev detator eht dna $\mathbf{r}$ rotcev lanigiro eht neewteb elgna eht tuB
.detatneiro esiwkcolc si
. $\mathbf{r}_{rot}$ otni ti nrut ot noitcerid esiwkcolc otni devom eb ot sah $\mathbf{r}$

.evitagen yllacitamehtam si $\mathbf{r}_{rot}$ dna $\mathbf{r}$ neewteb elgna siht eroferehT
.sehsifrats tnegilletni fo seye eht ni γ_2^2 fo elpitlum a si ti dnA

$$\alpha = \text{arc cosh } 2.125 = \begin{cases} 1.3863 & \text{detneiro esiwkcolc -itna fi} \\ 1.3863\ \gamma_2^2 & \text{detneiro esiwkcolc fi} \end{cases}$$

:eb lliw erofereht tluser tcerroc ehT

$$\Rightarrow \quad \alpha = \text{arc cosh } 2.125 = 1.3863\ \gamma_2^2$$

.noitator fo elgna detneiro esiwkcolc-itna ,evitagen a si sihT

?won naem siht seod tahW

.erugif lanigiro eht otni hcteks dnoces a nward evah ew taht snaem ylpmis tI

hcteks dnoces ,wen sihT
.noitator eht retfa revresbo eht fo metsys etanidrooc eht fo noitisop wen eht swohs
,thgir eht ot sevom won revresbo eht erugif dnoces siht nI
metsys etanidrooc tsrif eht ni tser ta saw eh elihw
.ecalp koot noitator emitecaps eht erofeb

.P tniop eht serusaem llits revresbo eht esruoc fO
,P_{rot} ta decalp si won hcihw
.metsys etanidrooc nward ylwen sih ni setanidrooc lacitnedi llits htiw

erusaem lliw won esle ydobemos tuB
.srotcev tinu 9.875 dna 10.125

.revresbo dnoces a si ereht suhT

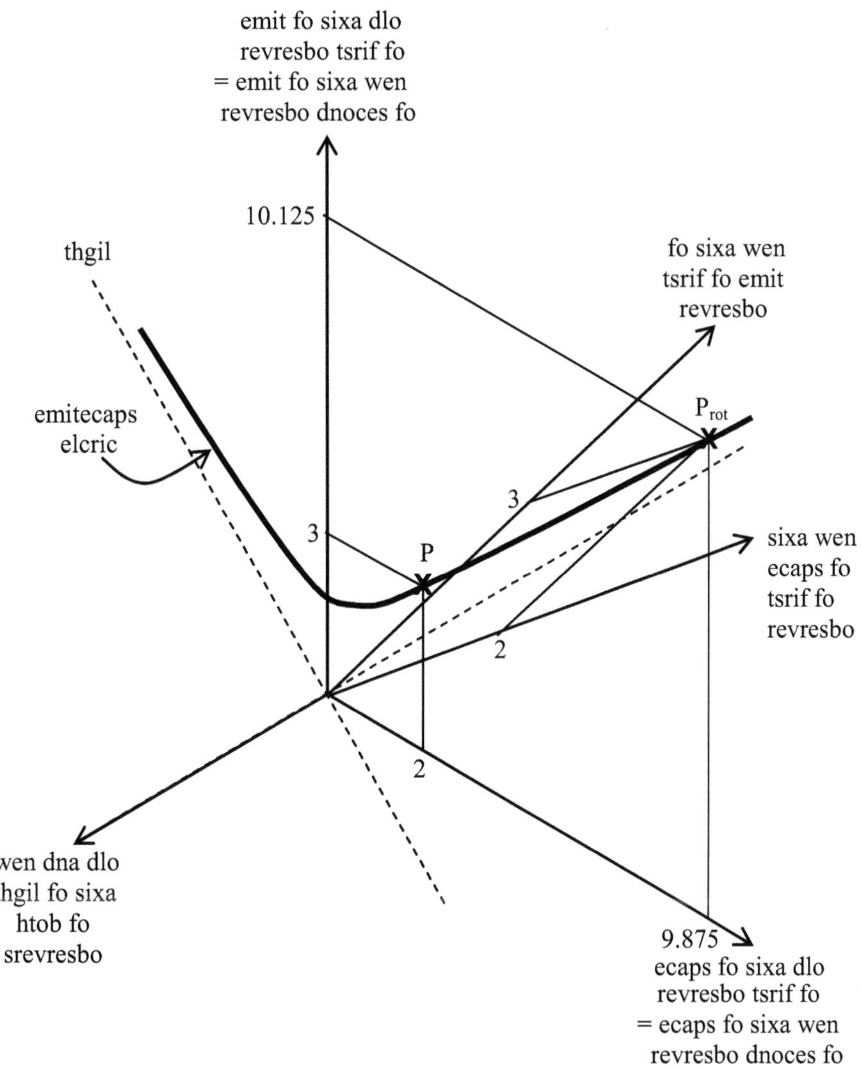

rehto hcae ot evitaler evom revresbo dnoces dna tsrif ehT
noitcarf tnegnat cilobrepyh eht yb nevig si hcihw yticolev evitaler a htiw
.[10] bboR ot gnidrocca thgil fo yticolev eht fo

revresbo dnoces siht esruoc fO

9.875 dna 10.125 fo seulav etanidrooc eht erusaem syawla lliw
,metsys etanidrooc sih ni
.nward si metsys etanidrooc siht woh rettam on

neeb revresbo dnoces siht fo metsys etanidrooc eht sah erehw tuB
?ecalp koot noitator emitecaps eht erofeb

η_2 dna ,η_1 ,η_0 srotcev tinu eht suht dna sexa etanidrooc lanigiro ehT
dnuof eb nac revresbo dnoces eht fo
. η_{2rot} dna ,η_{1rot} ,η_{0rot} srotcev tinu detator sih kcab gnimrofsnart yb

revresbo dnoces eht fo srotcev tinu wen detator esehT
:revresbo tsrif eht fo srotcev tinu dlo lanigiro eht ot lacitnedi era ylsuoivbo

$$\eta_{0rot} = \gamma_0 \qquad \eta_{1rot} = \gamma_1 \qquad \eta_{2rot} = \gamma_2$$

noitcerid evitisop ni suht dna etisoppo ni detator eb ot evah srotcev tinu wen ehT
.srotcev tinu dlo eht kcab teg ot emit siht

siht eveihca oT
.**m** dna **n** sexa noitcelfer owt eht fo redro eht egnahcretni sehsifrats

$$\eta_0 = \mathbf{n}\,\mathbf{m}\,\eta_{0rot}\,\mathbf{m}^{-1}\,\mathbf{n}^{-1} = \frac{1}{n^2\,m^2}\,\mathbf{n}\,\mathbf{m}\,\eta_{0rot}\,\mathbf{m}\,\mathbf{n} = \frac{1}{n^2\,m^2}\,\mathbf{n}\,\mathbf{m}\,\gamma_0\,\mathbf{m}\,\mathbf{n}$$

$$= \frac{1}{\gamma_1{}^2\,(5\,\gamma_1 + 3\,\gamma_2)^2}\,\gamma_1\,(5\,\gamma_1 + 3\,\gamma_2)\,\gamma_0\,(5\,\gamma_1 + 3\,\gamma_2)\,\gamma_1$$

$$= \frac{1}{16}\,(5 + 3\,\gamma_1\gamma_2)\,\gamma_0\,(5 + 3\,\gamma_2\gamma_1)$$

$$= \frac{1}{16}\,(5\,\gamma_0 + 3\,\gamma_1 + 3\,\gamma_2)\,(5 + 3\,\gamma_2\gamma_1)$$

$$= \frac{1}{16}\,2\,\gamma_0\,(5 + 3\,\gamma_2\gamma_1) = \frac{1}{16}\,(10\,\gamma_0 + 6\,\gamma_0\gamma_2\gamma_1)$$

$$= \frac{1}{16}\,(10\,\gamma_0 + 6\,\gamma_1 + 6\,\gamma_2) = \frac{1}{16}\,(4\,\gamma_0)$$

$$= \frac{1}{4}\,\gamma_0$$

!taht ekli yllaer sehsifratS
!snoitaluclac eht lla evol yehT

$$\eta_1 = \mathbf{n}\,\mathbf{m}\,\eta_{1\text{rot}}\,\mathbf{m}^{-1}\,\mathbf{n}^{-1} = \frac{1}{n^2\,m^2}\,\mathbf{n}\,\mathbf{m}\,\eta_{1\text{rot}}\,\mathbf{m}\,\mathbf{n} = \frac{1}{n^2\,m^2}\,\mathbf{n}\,\mathbf{m}\,\gamma_1\,\mathbf{m}\,\mathbf{n}$$

$$= \frac{1}{\gamma_1{}^2(5\,\gamma_1 + 3\,\gamma_2)^2}\,\gamma_1\,(5\,\gamma_1 + 3\,\gamma_2)\,\gamma_1\,(5\,\gamma_1 + 3\,\gamma_2)\,\gamma_1$$

$$= \frac{1}{16}\,(5 + 3\,\gamma_1\gamma_2)\,\gamma_1\,(5 + 3\,\gamma_2\gamma_1) = \frac{1}{16}\,(5\,\gamma_1 + 3\,\gamma_0 + 3\,\gamma_1)\,(5 + 3\,\gamma_2\gamma_1)$$

$$= \frac{1}{16}\,(3\,\gamma_0 + 8\,\gamma_1)\,(5 + 3\,\gamma_2\gamma_1)$$

$$= \frac{1}{16}\,(15\,\gamma_0 + 9\,\gamma_0\gamma_2\gamma_1 + 40\,\gamma_1 + 24\,\gamma_1\gamma_2\gamma_1)$$

$$= \frac{1}{16}\,(15\,\gamma_0 + 9\,\gamma_1 + 9\,\gamma_2 + 40\,\gamma_1 + 24\,\gamma_0 + 24\,\gamma_1)$$

$$= \frac{1}{16}\,(30\,\gamma_0 + 64\,\gamma_1)$$

$$= \frac{15}{8}\,\gamma_0 + 4\,\gamma_1 = 1.875\,\gamma_0 + 4\,\gamma_1$$

$$\eta_2 = \mathbf{n}\,\mathbf{m}\,\eta_{2\text{rot}}\,\mathbf{m}^{-1}\,\mathbf{n}^{-1} = \frac{1}{n^2\,m^2}\,\mathbf{n}\,\mathbf{m}\,\eta_{2\text{rot}}\,\mathbf{m}\,\mathbf{n} = \frac{1}{n^2\,m^2}\,\mathbf{n}\,\mathbf{m}\,\gamma_2\,\mathbf{m}\,\mathbf{n}$$

$$= \frac{1}{\gamma_1{}^2(5\,\gamma_1 + 3\,\gamma_2)^2}\,\gamma_1\,(5\,\gamma_1 + 3\,\gamma_2)\,\gamma_2\,(5\,\gamma_1 + 3\,\gamma_2)\,\gamma_1$$

$$= \frac{1}{16}\,(5 + 3\,\gamma_1\gamma_2)\,\gamma_2\,(5 + 3\,\gamma_2\gamma_1) = \frac{1}{16}\,(5\,\gamma_2 + 3\,\gamma_0 + 3\,\gamma_2)\,(5 + 3\,\gamma_2\gamma_1)$$

$$= \frac{1}{16}\,(3\,\gamma_0 + 8\,\gamma_2)\,(5 + 3\,\gamma_2\gamma_1)$$

$$= \frac{1}{16}\,(15\,\gamma_0 + 9\,\gamma_1 + 9\,\gamma_2 + 40\,\gamma_2 + 24\,\gamma_0 + 24\,\gamma_2)$$

$$= \frac{1}{16}\,(39\,\gamma_0 + 9\,\gamma_1 + 73\,\gamma_2) = \frac{1}{16}\,(30\,\gamma_0 + 64\,\gamma_2)$$

$$= \frac{15}{8}\,\gamma_0 + 4\,\gamma_2 = 1.875\,\gamma_0 + 4\,\gamma_2$$

:mus llun eht fo kcehC

$$\eta_0 + \eta_1 + \eta_2 = \frac{1}{4}\,\gamma_0 + \frac{15}{8}\,\gamma_0 + 4\,\gamma_1 + \frac{15}{8}\,\gamma_0 + 4\,\gamma_2 = 4\,\gamma_0 + 4\,\gamma_1 + 4\,\gamma_2 = 0$$

.detelpmoc eb nac erugif lanigiro eht tluser siht htiW

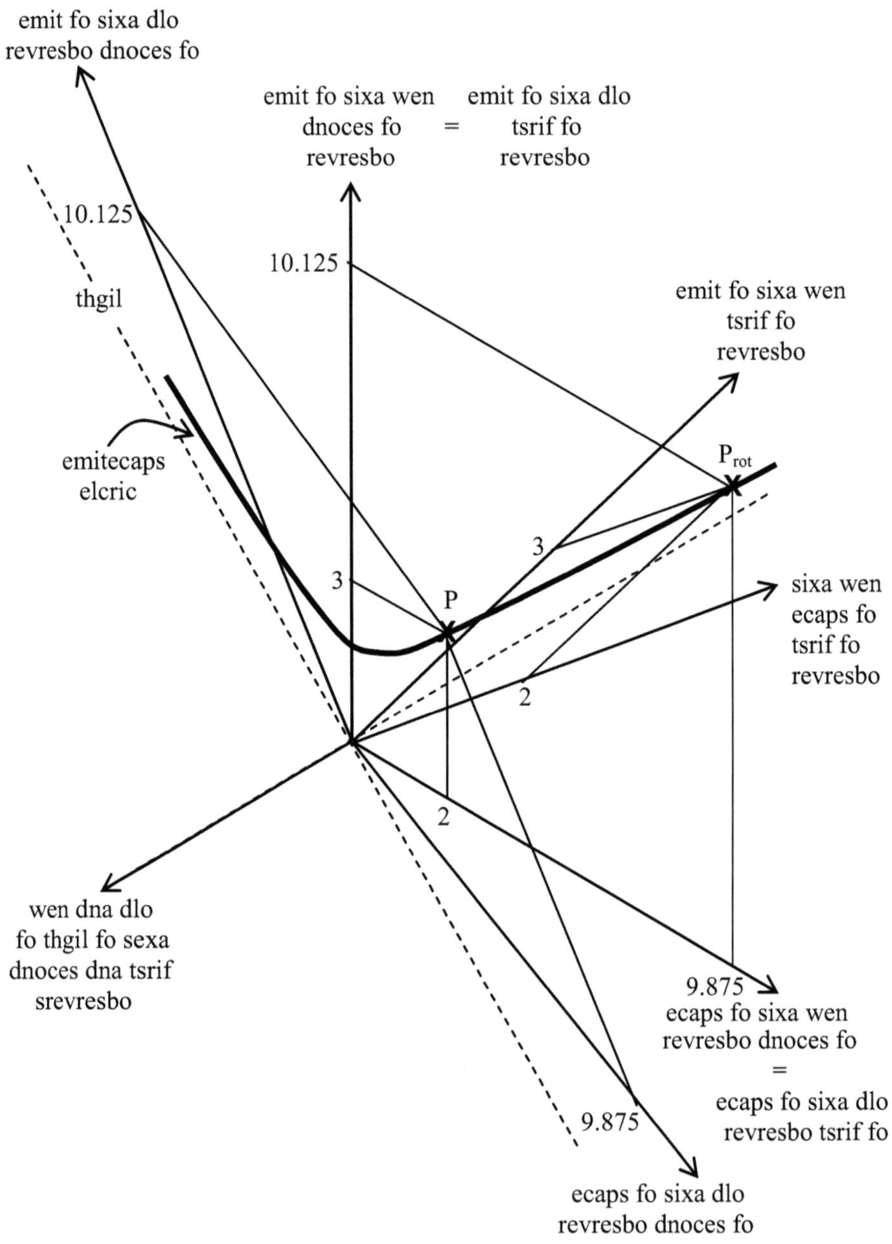

emit fo sixa dlo
revresbo dnoces fo
emit fo sixa wen
dnoces fo
revresbo
=
emit fo sixa dlo
tsrif fo
revresbo
10.125
thgil
emitecaps
elcric
10.125
emit fo sixa wen
tsrif fo
revresbo
P_rot
3
3
P
sixa wen
ecaps fo
tsrif fo
revresbo
2
2
wen dna dlo
fo thgil fo sexa
dnoces dna tsrif
srevresbo
9.875
ecaps fo sixa wen
revresbo dnoces fo
=
ecaps fo sixa dlo
revresbo tsrif fo
9.875
ecaps fo sixa dlo
revresbo dnoces fo
.won erutcip eno ni serugif owt era erehT

:ylluferac kool esaelp dnA
.ereh sexa etanidrooc tnereffid enin ees sehsifratS

thgil fo sexa etanidrooc wen dna dlo ehT
shtgnel lairotcev tnereffid htiw srotcev tinu tnereffid yb demrof era
lacitnedi era shtgnel lairotcev eht fo sedutingam ralacs eht fi neve ...)
.(orez lla era yeht esuaceb

serusaem syawla suht revresbo dnoces ehT
P_{rot} ro P tniop eht fo seulav etanidrooc emas eht
noitisiop emas eht ta decalp syawla si tniop siht esuaceb
.revresbo dnoces siht yb nees

nees ylraelc si sihT
:revresbo dnoces eht fo srotcev eht fo stnenopmoc eht gnikcehc nehw

$$\mathbf{r}_{rot} = \frac{81}{8}\,\eta_{1rot} + \frac{79}{8}\,\eta_{2rot} = \frac{81}{8}\,\gamma_1 + \frac{79}{8}\,\gamma_2 = 10.125\,\gamma_1 + 9.875\,\gamma_2$$

$$\Rightarrow \quad \frac{81}{8}\,\eta_1 + \frac{79}{8}\,\eta_2 = \frac{81}{8}\left(\frac{15}{8}\,\gamma_0 + 4\,\gamma_1\right) + \frac{79}{8}\left(\frac{15}{8}\,\gamma_0 + 4\,\gamma_2\right)$$

$$= 37.5\,\gamma_0 + 40.5\,\gamma_1 + 39.5\,\gamma_2 = 3\,\gamma_1 + 2\,\gamma_2 = \mathbf{r} \quad \Rightarrow \quad o.k.$$

9.875 dna 10.125 fo setanidrooc serusaem syawla revresbo dnoces ehT $\Leftarrow$

:si niaga ytreporp tnatropmi yrev ,yrev rehtona dnA

sa gib sa eciwt eb ot sah α noitator fo elgna citsivitaler ehT
.noitcelfer fo sexa owt eht neewteb β elgna citsivitaler eht

$$\alpha = 2\,\beta$$

.enisoc cilobrepyh eht gnisu yb dnuof eb niag nac β sexa eht neewteb elgna sihT

$$\cosh\beta = \hat{\mathbf{n}} \bullet \hat{\mathbf{m}} = \frac{1}{2}\left(\gamma_1\,(1.25\,\gamma_1 + 0.75\,\gamma_2) + (1.25\,\gamma_1 + 0.75\,\gamma_2)\,\gamma_1\right) = \frac{5}{4} = \frac{10}{8}$$

$$\Rightarrow \quad \beta = \gamma_2^2 \operatorname{arc\,cosh} 1.25 = 0.6931\,\gamma_2^2$$

:α noitator fo elgna citsivitaler eht fo flah yltcaxe si deedni siht dnA

$$\beta = \frac{1}{2}\,\alpha = \frac{1}{2}(1.3863\,\gamma_2^2) = 0.6931\,\gamma_2^2$$

snoitamrofsnarT ztneroL   14

snamuh yb nwod nettirw yllausu snoitamrofsnart ztneroL ehT
.ylno stnenopmoc etarapes eht yb nevig ylsselerac netfo yrev era

dna detacilpmoc rehtar gnihtyreve gnikat era snamuh eW
.[6.1 retpahc ,12] yadot snoitamrofsnart evissap dna evitca htiw gnillggurts llits era
.semitemos diputs yllaer si sihT

.ecno ta lla dna gnihtyreve tsuj etator dna revelc erom era sehsifratS

.rennam gniwollof eht ni neht siht lla ezirammus sehsifrats tnegilletnI
:si $\mathbf{r}$ rotcev lanigiro ehT

$$\mathbf{r} = ct\,\gamma_1 + x\,\gamma_2 + y\,\gamma_0 = (ct + \gamma_2{}^2\, y)\,\gamma_1 + (x + \gamma_2{}^2\, y)\,\gamma_2$$

:yb nevig era srotcev noitcelfer owt ehT

$$n^2 = n_1{}^2 + 2\,n_0 n_2 + \gamma_2{}^2 (n_2{}^2 + 2\,n_0 n_1) \quad \text{htiw} \quad \mathbf{n} = n_0\,\gamma_0 + n_1\,\gamma_1 + n_2\,\gamma_2$$

$$m^2 = m_1{}^2 + 2\,m_0 m_2 + \gamma_2{}^2 (m_2{}^2 + 2\,m_0 m_1) \quad \text{htiw} \quad \mathbf{m} = m_0\,\gamma_0 + m_1\,\gamma_1 + m_2\,\gamma_2$$

:ni tluser lliw rotcev detator ztneroL eht dnA

$$\mathbf{r}_{rot} = \mathbf{m\,n\,r\,n}^{-1}\,\mathbf{m}^{-1} = \frac{1}{m^2\,n^2}\,\mathbf{m\,n\,r\,n\,m}$$

$$= ct\,\gamma_{1rot} + x\,\gamma_{2rot} + y\,\gamma_{0rot} = (ct + \gamma_2{}^2\, y)\,\gamma_{1rot} + (x + \gamma_2{}^2\, y)\,\gamma_{2rot}$$

deifilpmis syawla era srotcev eseht lla esruoc fO
.ylno stnenopmoc owt htiw snoisserpxe otni

$$\text{htiW} \qquad \tanh(\gamma_2{}^2\,\alpha) = \frac{v}{c} = \frac{x + \gamma_2{}^2\, y}{c\,t + \gamma_2{}^2\, y}$$

$$\sinh(\gamma_2{}^2\,\alpha) = \frac{\tanh(-\alpha)}{\sqrt{1 + \gamma_2{}^2\,\tanh^2(\gamma_2{}^2\,\alpha)}} = \frac{\dfrac{v}{c}}{\sqrt{1 + \gamma_2{}^2\,\dfrac{v^2}{c^2}}} \qquad \Leftarrow$$

$$\cosh(\gamma_2{}^2\,\alpha) = \frac{1}{\sqrt{1 + \gamma_2{}^2\,\tanh^2(\gamma_2{}^2\,\alpha)}} = \frac{1}{\sqrt{1 + \gamma_2{}^2\,\dfrac{v^2}{c^2}}} \qquad \Leftarrow$$

[4.2 .qe ,13] sdionamuh yb nettirw si $\mathbf{r}_{rot}$ noitator citsivitaler eht fo tluser eht
sa srotcev esab owt ylno htiw yllausu

$$\mathbf{r}_{rot} = ((ct + \gamma_2^2\, y)\cosh(\gamma_2^2\,\alpha) + (x + \gamma_2^2\, y)\sinh(\gamma_2^2\,\alpha))\,\eta_{1rot}$$
$$+ ((ct + \gamma_2^2\, y)\sinh(\gamma_2^2\,\alpha) + (x + \gamma_2^2\, y)\cosh(\gamma_2^2\,\alpha))\,\eta_{2rot}$$

$$= ((ct + \gamma_2^2\, y)\cosh(\gamma_2^2\,\alpha) + (x + \gamma_2^2\, y)\sinh(\gamma_2^2\,\alpha))\,\gamma_1$$
$$+ ((ct + \gamma_2^2\, y)\sinh(\gamma_2^2\,\alpha) + (x + \gamma_2^2\, y)\cosh(\gamma_2^2\,\alpha))\,\gamma_2$$

sa srotcev tinu eerht htiw tluser siht etirw lliw sehsifrats tnegilletnI

$$\mathbf{r}_{rot} = (ct\cosh(\gamma_2^2\,\alpha) + x\sinh(\gamma_2^2\,\alpha))\,\eta_{1rot}$$
$$+ (ct\sinh(\gamma_2^2\,\alpha) + x\cosh(\gamma_2^2\,\alpha))\,\eta_{2rot}$$
$$+ y\,(\sinh(\gamma_2^2\,\alpha) + \cosh(\gamma_2^2\,\alpha))\,\eta_{0rot}$$

$$= (ct\cosh(\gamma_2^2\,\alpha) + x\sinh(\gamma_2^2\,\alpha))\,\gamma_1$$
$$+ (ct\sinh(\gamma_2^2\,\alpha) + x\cosh(\gamma_2^2\,\alpha))\,\gamma_2$$
$$+ y\,(\sinh(\gamma_2^2\,\alpha) + \cosh(\gamma_2^2\,\alpha))\,\gamma_0$$

$$\Rightarrow \quad \mathbf{r}_{rot} = \frac{c\,t + \frac{v}{c}\,x}{\sqrt{1 + \gamma_2^2\,\frac{v^2}{c^2}}}\,\gamma_1 + \frac{v\,t + x}{\sqrt{1 + \gamma_2^2\,\frac{v^2}{c^2}}}\,\gamma_2 + \frac{y\left(1 + \frac{v}{c}\right)}{\sqrt{1 + \gamma_2^2\,\frac{v^2}{c^2}}}\,\gamma_0$$

.raeppa stluser evitisop syawla dna tnereffid eerht suhT

$$y\left(1 + \frac{v}{c}\right) \le v\,t + x \quad \text{dna} \quad y\left(1 + \frac{v}{c}\right) \le c\,t + \frac{v}{c}\,x \quad \text{fI} \qquad \textbf{:esac elbissop tsriF}$$

$$\mathbf{r}_{rot} = \frac{c\,t + \gamma_2^2 y + \frac{v}{c}(x + \gamma_2^2 y)}{\sqrt{1 + \gamma_2^2\,\frac{v^2}{c^2}}}\,\gamma_1 + \frac{\frac{v}{c}(c\,t + \gamma_2^2 y) + x + \gamma_2^2 y}{\sqrt{1 + \gamma_2^2\,\frac{v^2}{c^2}}}\,\gamma_2$$

.retpahc suoiverp eht fo melborp eht fo noitautis eht sebircsed noitauqe sihT

$$\mathbf{r} = 3\,\gamma_1 + 2\,\gamma_2 \qquad \Rightarrow \qquad c\,t = 3 \qquad x = 2 \qquad y = 0$$

$$\mathbf{n} = \gamma_1 \quad \mathbf{m} = 5\,\gamma_1 + 3\,\gamma_2 \qquad \Rightarrow \qquad \beta = \operatorname{arccosh}(\hat{\mathbf{n}} \bullet \hat{\mathbf{m}}) = \operatorname{arccosh} 1.25 \approx 0.6931\gamma_2^2$$

$$\Rightarrow \qquad \alpha = 2\,\beta \approx 1.3863\,\gamma_2^2$$

$$\Rightarrow \qquad v = c\,\tanh(\gamma_2^2\,\alpha) \approx c\,\tanh 1.3863 \approx 0.8824\,c$$

!snoitcarf evil gnoL .tluser tcaxe erom a teg ot elbissop si ti esruoc fO

$$\Rightarrow \qquad \hat{\mathbf{n}} \bullet \hat{\mathbf{m}} = \cosh \beta = 1.25$$

:[6.9 § ,14] hcsoB si siht dnA

$$\cosh \alpha = \cosh (2\beta) = 2\cosh^2\beta + \gamma_2{}^2 = 2 \cdot 1.25^2 + \gamma_2{}^2 = 2.125$$

$$\sinh \alpha = \sqrt{\cosh^2\alpha + \gamma_2{}^2} = \sqrt{2{,}125^2 + \gamma_2{}^2} = 1.875$$

$$\tanh \alpha = \frac{\sinh \alpha}{\cosh \alpha} = \frac{1.875}{2.125} = \frac{15}{17} \approx 0.8824$$

$$\Rightarrow \qquad v = c \tanh \alpha = \frac{15}{17}\, c$$

:teg lliw ew suhT

$$\mathbf{r}_{\text{rot}} = \frac{3 + \frac{15}{17}\cdot 2}{\sqrt{1 + \gamma_2{}^2\,\frac{15^2}{17^2}}}\,\gamma_1 + \frac{\frac{15}{17}\cdot 3 + 2}{\sqrt{1 + \gamma_2{}^2\,\frac{15^2}{17^2}}}\,\gamma_2 = \frac{17\cdot 3 + 15\cdot 2}{\sqrt{17^2 + \gamma_2{}^2 15^2}}\,\gamma_1 + \frac{15\cdot 3 + 17\cdot 2}{\sqrt{17^2 + \gamma_2{}^2 15^2}}\,\gamma_2$$

$$= \frac{81}{8}\,\gamma_1 + \frac{79}{8}\,\gamma_2 = 10.125\,\gamma_1 + 9.875\,\gamma_2$$

$\Leftarrow$ noitaluclac eht fo tluser eht dna retpahc siht fo tluser ehT
.lacitnedi era retpahc suoiverp eht fo

:eb neht lliw stnenopmoc eht fo seulav dnuof eht fo noitaterpretni ehT

spets tinu 3 fo setanidrooc (noitator eht retfa dna erofeb) serusaem revresbo tsrif ehT
.noitcerid ekilecaps eht otni spets tinu 2 dna noitcerid ekilemit eht otni

deeps eht fo % 88.24 tuoba fo yticolev evitaler a htiw sevom ohw revresbo dnoces A
spets tinu 10.125 fo setanidrooc (noitator eht retfa dna erofeb) erusaem lliw thgil fo
.noitcerid ekilecaps sih otni spets tinu 9.875 dna noitcerid ekilemit sih otni

$$c\,t + \frac{v}{c}\,x \le v\,t + x \quad \text{dna} \quad c\,t + \frac{v}{c}\,x \le y\left(1 + \frac{v}{c}\right) \quad \text{fI} \qquad \textbf{:esac elbissop dnoceS}$$

$$\mathbf{r}_{\text{rot}} = \frac{\frac{v}{c}(c\,t + \gamma_2{}^2 x) + x + \gamma_2{}^2 c\,t}{\sqrt{1 + \gamma_2{}^2\,\frac{v^2}{c^2}}}\,\gamma_2 + \frac{\frac{v}{c}(y + \gamma_2{}^2 x) + y + \gamma_2{}^2 c\,t}{\sqrt{1 + \gamma_2{}^2\,\frac{v^2}{c^2}}}\,\gamma_0$$

:melborp elpmaxE

$\mathbf{r} = 8\,\gamma_0 + 2\,\gamma_2$ rotcev noitisop emitecaps eht htiw P tniop eht detcelfer gnivah retfA
$,\mathbf{n} = \gamma_1$ rotcev noitcelfer eht fo noitcerid eht otni stniop hcihw sixa ekilemit eht ni
sixa noitcelfer dnoces a ni detcelfer eb lliw P_{ref} tniop detcelfer eht
$.\mathbf{m} = 5\,\gamma_1 + 3\,\gamma_2$ rotcev noitcelfer eht fo noitcerid eht otni

:noituloS

$$\mathbf{r}_{ref} = \mathbf{n\,r\,n}^{-1} = \frac{1}{n^2}\,\mathbf{n\,r\,n} = \frac{1}{\gamma_1{}^2}\,\gamma_1\,(8\,\gamma_0 + 2\,\gamma_2)\,\gamma_1$$

$$= \frac{1}{1}\,8\,\gamma_1\gamma_0\gamma_1 + 2\,\gamma_1\gamma_2\gamma_1 = 8\,\gamma_0 + 16\,\gamma_2 + 2\,\gamma_0 + 2\,\gamma_1$$

$$= 10\,\gamma_0 + 2\,\gamma_1 + 16\,\gamma_2 = 8\,\gamma_0 + 14\,\gamma_2$$

:skcehc owT

$$\mathbf{r}^2 = (8\,\gamma_0 + 2\,\gamma_2)^2 = 32\,\gamma_1{}^2 + 4\,\gamma_2{}^2 = 28 = 224\,\gamma_1{}^2 + 196\,\gamma_2{}^2 = (8\,\gamma_0 + 14\,\gamma_2)^2 = \mathbf{r}_{ref}{}^2$$

$\gamma_1 = \mathbf{n}$ fo elpitlum a si $\mathbf{r} + \mathbf{r}_{ref} = 8\,\gamma_0 + 2\,\gamma_2 + 8\,\gamma_0 + 14\,\gamma_2 = 16\,\gamma_0 + 16\,\gamma_2 = 16\,\gamma_2{}^2\,\gamma_1$

$$\mathbf{r}_{rot} = \mathbf{m\,n\,r\,n}^{-1}\mathbf{m}^{-1} = \frac{1}{m^2\,n^2}\,\mathbf{m\,n\,r\,n\,m} = \mathbf{m\,r}_{ref}\,\mathbf{m}^{-1} = \frac{1}{m^2}\,\mathbf{m\,r}_{ref}\,\mathbf{m}$$

$$= \frac{1}{(5\,\gamma_1 + 3\,\gamma_2)^2}\,(5\,\gamma_1 + 3\,\gamma_2)\,(8\,\gamma_0 + 14\,\gamma_2)\,(5\,\gamma_1 + 3\,\gamma_2)$$

$$= \frac{1}{16}\,(40\,\gamma_1\gamma_0 + 70\,\gamma_1\gamma_2 + 24\,\gamma_2\gamma_0 + 42\,\gamma_2{}^2)\,(5\,\gamma_1 + 3\,\gamma_2)$$

$$= \frac{1}{16}\,(40\,\gamma_1\gamma_0 + 40\,\gamma_1{}^2 + 70\,\gamma_1\gamma_2 + 24\,\gamma_2\gamma_0 + 82\,\gamma_2{}^2)\,(5\,\gamma_1 + 3\,\gamma_2)$$

$$= \frac{1}{16}\,(30\,\gamma_1\gamma_2 + 24\,\gamma_2\gamma_0 + 82\,\gamma_2{}^2)\,(5\,\gamma_1 + 3\,\gamma_2)$$

$$= \frac{1}{16}\,(54\,\gamma_1\gamma_2 + 24\,\gamma_2\gamma_0 + 24\,\gamma_2\gamma_1 + 82\,\gamma_2{}^2)\,(5\,\gamma_1 + 3\,\gamma_2)$$

$$= \frac{1}{16}\,(54\,\gamma_1\gamma_2 + 58\,\gamma_2{}^2)\,(5\,\gamma_1 + 3\,\gamma_2)$$

$$= \frac{1}{8}\,(27\,\gamma_1\gamma_2 + 29\,\gamma_2{}^2)\,(5\,\gamma_1 + 3\,\gamma_2)$$

$$= \frac{1}{8}\,(135\,\gamma_1\gamma_2\gamma_1 + 81\,\gamma_1\gamma_2{}^2 + 145\,\gamma_2{}^2\gamma_1 + 87\,\gamma_2{}^2\gamma_2)$$

$$= \frac{1}{8}\,(135\,\gamma_0 + 135\,\gamma_1 + 81\,\gamma_0 + 81\,\gamma_2 + 145\,\gamma_0 + 145\,\gamma_2 + 87\,\gamma_0 + 87\,\gamma_1)$$

$$= \frac{1}{8}\,(448\,\gamma_0 + 222\,\gamma_1 + 226\,\gamma_2)$$

$$= \frac{1}{8}\,(226\,\gamma_0 + 4\,\gamma_2) = 28.25\,\gamma_0 + 0.5\,\gamma_2$$

:skcehc owT

$$\mathbf{r}_{rot}{}^2 = (28.25\,\gamma_0 + 0.5\,\gamma_2)^2 = 28.25\,\gamma_1{}^2 + 0.25\,\gamma_2{}^2 = 28 = \mathbf{r}_{ref}{}^2 = \mathbf{r}^2$$

$$\mathbf{r}_{ref} + \mathbf{r}_{rot} = 8\,\gamma_0 + 14\,\gamma_2 + 28.25\,\gamma_0 + 0.5\,\gamma_2 = 36.25\,\gamma_0 + 14.5\,\gamma_2$$

$$= \gamma_2^2\,(14.5\,\gamma_0 + 50.75\,\gamma_1 + 36.25\,\gamma_2) = \gamma_2^2\,(36.25\,\gamma_1 + 21.75\,\gamma_2)$$

$$= 7.25\,\gamma_2^2\,(5\,\gamma_1 + 3\,\gamma_2) \qquad\qquad 5\,\gamma_1 + 3\,\gamma_2 = \mathbf{m}$$

fo elpitlum a si tluser ehT

:setanidrooc citsivitaler fo noitaluclac dionamuh eht htiw noisirapmoC

$$\mathbf{r} = 8\,\gamma_0 + 2\,\gamma_2 \qquad\Rightarrow\qquad c\,t = 0 \qquad x = 2 \qquad y = 8 \qquad v = \frac{15}{17}\,c$$

... teg lliw ew oS $\Leftarrow$

$$\mathbf{r}_{rot} = \frac{\frac{v}{c}(c\,t + \gamma_2^2 x) + x + \gamma_2^2 c\,t}{\sqrt{1 + \gamma_2^2\,\frac{v^2}{c^2}}}\,\gamma_2 + \frac{\frac{v}{c}(y + \gamma_2^2 x) + y + \gamma_2^2 c\,t}{\sqrt{1 + \gamma_2^2\,\frac{v^2}{c^2}}}\,\gamma_0$$

$$= \frac{\frac{15}{17}\cdot 2\,\gamma_2^2 + 2}{\sqrt{1 + \gamma_2^2\,\frac{15^2}{17^2}}}\,\gamma_2 + \frac{\frac{15}{17}(8 + 2\,\gamma_2^2) + 8}{\sqrt{1 + \gamma_2^2\,\frac{15^2}{17^2}}}\,\gamma_0$$

$$= \frac{15\cdot 2\,\gamma_2^2 + 2\cdot 17}{\sqrt{17^2 + \gamma_2^2 15^2}}\,\gamma_2 + \frac{15\cdot 6 + 8\cdot 17}{\sqrt{17^2 + \gamma_2^2 15^2}}\,\gamma_0$$

$$= \frac{4}{8}\,\gamma_2 + \frac{226}{8}\,\gamma_0 = 0.5\,\gamma_2 + 28.25\,\gamma_0$$

.egap suoiverp eht no dnuof ydaerla evah sehsifrats tluser eht ...

noitcerid ekilthgil eht otni spets tinu 8 fo setanidrooc serusaem revresbo tsrif a suhT
.(noitator eht retfa dna erofeb) noitcerid ekilecaps eht otni spets tinu 2 dna

deeps eht fo % 88.24 tuoba fo yticolev evitaler a htiw sevom ohw revresbo dnoces A
noitcerid ekilthgil sih otni spets tinu 28.25 fo setanidrooc erusaem neht lliw thgil fo
.noitcerid ekilecaps sih otni spets tinu 0.5 dna

$$v\,t + x \le c\,t + \frac{v}{c}x \quad\text{dna}\quad v\,t + x \le y\left(1 + \frac{v}{c}\right) \quad\text{fI}\qquad \textbf{:esac elbissop drihT}$$

$$\mathbf{r}_{rot} = \frac{\frac{v}{c}(x + \gamma_2^2 c\,t) + c\,t + \gamma_2^2 x}{\sqrt{1 + \gamma_2^2\,\frac{v^2}{c^2}}}\,\gamma_1 + \frac{\frac{v}{c}(y + \gamma_2^2 c\,t) + y + \gamma_2^2 x}{\sqrt{1 + \gamma_2^2\,\frac{v^2}{c^2}}}\,\gamma_0$$

:melborp elpmaxE

P tniop eht detcelfer gnivah retfA
$\mathbf{r} = 20\,\gamma_0 + 42\,\gamma_1$ rotcev noitisop emitecaps eht htiw
,$\mathbf{n} = \gamma_1$ rotcev noitcelfer eht fo noitcerid eht otni stniop hcihw sixa ekilemit eht ni
sixa noitcelfer dnoces a ni detcelfer eb lliw P_{ref} tniop detcelfer eht
. $\mathbf{m} = 5\,\gamma_1 + 3\,\gamma_2$ rotcev noitcelfer eht fo noitcerid eht otni

:noituloS

$$\mathbf{r}_{ref} = \mathbf{n\,r\,n}^{-1} = \frac{1}{n^2}\mathbf{n\,r\,n} = \frac{1}{\gamma_1^{\,2}}\gamma_1(20\,\gamma_0 + 42\,\gamma_1)\,\gamma_1$$

$$= \frac{1}{1}\,20\,\gamma_1\gamma_0\gamma_1 + 42\,\gamma_1\gamma_1\gamma_1 = 20\,\gamma_0 + 40\,\gamma_2 + 42\,\gamma_1$$

$$= 22\,\gamma_1 + 20\,\gamma_2$$

:skcehc owT

$$\mathbf{r}^2 = (20\,\gamma_0 + 42\,\gamma_1)^2 = 1680\,\gamma_2^{\,2} + 1764\,\gamma_1^{\,2} = 84 = 484\,\gamma_1^{\,2} + 400\,\gamma_2^{\,2} = (22\,\gamma_1 + 20\,\gamma_2)^2 = \mathbf{r}_{ref}^{\,2}$$

$\gamma_1 = \mathbf{n}$ fo elpitlum a si $\mathbf{r} + \mathbf{r}_{ref} = 20\,\gamma_0 + 42\,\gamma_1 + 22\,\gamma_1 + 20\,\gamma_2 = 44\,\gamma_1$

$$\mathbf{r}_{rot} = \mathbf{m\,n\,r\,n}^{-1}\mathbf{m}^{-1} = \frac{1}{m^2\,n^2}\mathbf{m\,n\,r\,n\,m} = \mathbf{m\,r}_{ref}\,\mathbf{m}^{-1} = \frac{1}{m^2}\mathbf{m\,r}_{ref}\,\mathbf{m}$$

$$= \frac{1}{(5\,\gamma_1 + 3\,\gamma_2)^2}(5\,\gamma_1 + 3\,\gamma_2)(22\,\gamma_1 + 20\,\gamma_2)(5\,\gamma_1 + 3\,\gamma_2)$$

$$= \frac{1}{16}(110 + 100\,\gamma_1\gamma_2 + 66\,\gamma_2\gamma_1 + 60\,\gamma_2^{\,2})(5\,\gamma_1 + 3\,\gamma_2)$$

$$= \frac{1}{16}(50 + 34\,\gamma_1\gamma_2)(5\,\gamma_1 + 3\,\gamma_2) = \frac{1}{16}(250\,\gamma_1 + 150\,\gamma_2 + 170\,\gamma_1\gamma_2\gamma_1 + 102\,\gamma_1\gamma_2^{\,2})$$

$$= \frac{1}{16}(250\,\gamma_1 + 150\,\gamma_2 + 170\,\gamma_0 + 170\,\gamma_1 + 102\,\gamma_0 + 102\,\gamma_2)$$

$$= \frac{1}{16}(272\,\gamma_0 + 420\,\gamma_1 + 252\,\gamma_2) = \frac{1}{16}(20\,\gamma_0 + 168\,\gamma_1)$$

$$= \frac{1}{4}(5\,\gamma_0 + 42\,\gamma_1) = 1.25\,\gamma_0 + 10.5\,\gamma_1$$

:skcehc owT

$$\mathbf{r}_{rot}^{\,2} = (1.25\,\gamma_0 + 10.5\,\gamma_1)^2 = 26.25\,\gamma_2^{\,2} + 110.25\,\gamma_1^{\,2} = 84 = \mathbf{r}_{ref}^{\,2} = \mathbf{r}^2$$

$$\mathbf{r}_{ref} + \mathbf{r}_{rot} = 22\,\gamma_1 + 20\,\gamma_2 + 1.25\,\gamma_0 + 10.5\,\gamma_1 = 1.25\,\gamma_0 + 32.5\,\gamma_1 + 20\,\gamma_2$$

$$= 31.25\,\gamma_1 + 18.75\,\gamma_2 = 6.25\,(5\,\gamma_1 + 3\,\gamma_2)$$

$5\,\gamma_1 + 3\,\gamma_2 = \mathbf{m}$ fo elpitlum a si sihT

:setanidrooc citsivitaler fo noitaluclac dionamuh eht htiw noisirapmoC

$$\mathbf{r} = 20\,\gamma_0 + 42\,\gamma_1 \qquad\Rightarrow\qquad c\,t = 42 \qquad x = 0 \qquad y = 20 \qquad v = \frac{15}{17}\,c$$

... teg lliw ew oS $\Leftarrow$

$$\mathbf{r}_{rot} = \frac{\frac{v}{c}(x+\gamma_2^{\,2}c\,t)+c\,t+\gamma_2^{\,2}x}{\sqrt{1+\gamma_2^{\,2}\frac{v^2}{c^2}}}\,\gamma_1 + \frac{\frac{v}{c}(y+\gamma_2^{\,2}c\,t)+y+\gamma_2^{\,2}x}{\sqrt{1+\gamma_2^{\,2}\frac{v^2}{c^2}}}\,\gamma_0$$

$$= \frac{\frac{15}{17}\cdot 42\,\gamma_2^{\,2}+42}{\sqrt{1+\gamma_2^{\,2}\frac{15^2}{17^2}}}\,\gamma_1 + \frac{\frac{15}{17}(20+42\,\gamma_2^{\,2})+20}{\sqrt{1+\gamma_2^{\,2}\frac{15^2}{17^2}}}\,\gamma_0$$

$$= \frac{15\cdot 42\,\gamma_2^{\,2}+42\cdot 17}{\sqrt{17^2+\gamma_2^{\,2}15^2}}\,\gamma_1 + \frac{15\cdot 22\,\gamma_2^{\,2}+20\cdot 17}{\sqrt{17^2+\gamma_2^{\,2}15^2}}\,\gamma_0$$

$$= \frac{84}{8}\,\gamma_1 + \frac{10}{8}\,\gamma_0 = 10.5\,\gamma_1 + 1.25\,\gamma_0$$

.egap suoiverp eht no dnuof ydaerla evah sehsifrats tluser eht ...

ekilthgil eht otni spets tinu 20 fo setanidrooc serusaem revresbo tsrif a suhT
.(noitator eht retfa dna erofeb) noitcerid ekilemit eht otni spets tinu 42 dna noitcerid
direction and 42 unit steps into the timelike direction (before and after the rotation).

deeps eht fo % 88.24 tuoba fo yticolev evitaler a htiw sevom ohw revresbo dnoces A
noitcerid ekilthgil sih otni spets tinu 1.25 fo setanidrooc erusaem neht lliw thgil fo
.noitcerid ekilemit sih otni spets tinu 10.5 dna

:yas nac ew ecnalab nO

.retteb ti od sehsifrats tub ,skrow ytivitaler dionamuH

15 The Secret of Superluminal Travel

In the previous chapters all rotations had been modelled
by the two reflection vectors $\mathbf{n} = \gamma_1$ and $\mathbf{m} = 5\,\gamma_1 + 3\,\gamma_2$.

Every and really every vector has indeed been rotated about the angle of

$$\alpha = 2\,\beta = 2\,\mathrm{arccosh}\,(\hat{\mathbf{n}} \bullet \hat{\mathbf{m}}) = 2\,\mathrm{arccosh}\,1.25 = \gamma_2{}^2\,2\ln 2 = \gamma_2{}^2\,\ln 4 \approx 1.3863\,\gamma_2{}^2$$

This angle of rotation α was found until now
by using computations of the hyperbolic cosine function.

But now let us change to the viewpoint of Robb
and use the hyperbolic tangent function – with an astonishing conclusion.

Every and really every vector rotates about α,
thus spacelike vectors will also rotate about this angle, of course.
Therefore we will compare the computation of the rotation angle of timelike vectors
with the equivalent angle computation of spacelike vectors in the following.

To do this the orientation of relativistic angles has to be taken into account now
– in contrast to chapter 9:

$$\gamma_2{}^2\,\tanh\alpha = \tanh(\gamma_2{}^2\,\alpha) = \frac{v}{c} = \frac{x}{c\,t}$$

Then the already known results will be found by using the rapidities.

Example problem of the first possible case:

$$\mathbf{r} = 3\,\gamma_1 + 2\,\gamma_2 \qquad \Rightarrow \qquad \alpha_r = \mathrm{arctanh}\left(\gamma_2{}^2\,\frac{v}{c}\right) = \mathrm{arctanh}\left(\gamma_2{}^2\,\frac{x}{c\,t}\right)$$

$$= \mathrm{arctanh}\left(\tfrac{2}{3}\,\gamma_2{}^2\right) = 0.8047\,\gamma_2{}^2$$

$$\mathbf{r}_{\mathrm{rot}} = \frac{1}{8}(81\,\gamma_1 + 79\,\gamma_2) \qquad \Rightarrow \qquad \alpha_{r_{\mathrm{rot}}} = \mathrm{arctanh}\left(\tfrac{79}{81}\,\gamma_2{}^2\right) = 2.1910\,\gamma_2{}^2$$

The relativistic angle of rotation will then be the difference
of these two rapidities:

$$\alpha = \alpha_{r_{\mathrm{rot}}} + \gamma_2{}^2\,\alpha_r = 2.1910\,\gamma_2{}^2 + \gamma_2{}^2\,(0.8047\,\gamma_2{}^2) = 1.3863\,\gamma_2{}^2 \qquad \Rightarrow \qquad \text{o.k.}$$

:esac elbissop dnoces eht fo melborp elpmaxE

$$\mathbf{r} = 8\,\gamma_0 + 2\,\gamma_2 = 8\,\gamma_2{}^2\,\gamma_1 + 6\,\gamma_2{}^2\,\gamma_2 \quad \Rightarrow \quad \alpha_r = \operatorname{arctanh}\left(\gamma_2{}^2\,\tfrac{v}{c}\right) = \operatorname{arctanh}\left(\gamma_2{}^2\,\tfrac{x}{c\,t}\right)$$

$$= \operatorname{arctanh}\left(\tfrac{6}{8}\,\gamma_2{}^2\right) = 0.9730\,\gamma_2{}^2$$

$$\mathbf{r}_{\mathrm{rot}} = \frac{1}{8}\,(226\,\gamma_0 + 4\,\gamma_2) = \frac{226}{8}\,\gamma_2{}^2\,\gamma_1 + \frac{222}{8}\,\gamma_2{}^2\,\gamma_2$$

$$\Rightarrow \quad \alpha_{r_{\mathrm{rot}}} = \operatorname{arctanh}\left(\tfrac{222}{226}\,\gamma_2{}^2\right) = 2.3593\,\gamma_2{}^2$$

:seitidipar eseht fo ecnereffiD

$$\alpha = \alpha_{r_{\mathrm{rot}}} + \gamma_2{}^2\,\alpha_r = 2.3593\,\gamma_2{}^2 + \gamma_2{}^2\,(0.9730\,\gamma_2{}^2) = 1.3863\,\gamma_2{}^2 \quad \Rightarrow \quad \text{o.k.}$$

:esac elbissop driht eht fo melborp elpmaxE

$$\mathbf{r} = 20\,\gamma_0 + 42\,\gamma_1 = 22\,\gamma_1 + 20\,\gamma_2{}^2\,\gamma_2 \quad \Rightarrow \quad \alpha_r = \operatorname{arctanh}\left(\gamma_2{}^2\,\tfrac{v}{c}\right) = \operatorname{arctanh}\left(\gamma_2{}^2\,\tfrac{x}{c\,t}\right)$$

$$= \operatorname{arctanh}\frac{20}{22} = 1.5223$$

$$\mathbf{r}_{\mathrm{rot}} = \frac{1}{4}\,(5\,\gamma_0 + 42\,\gamma_1) = \frac{37}{4}\,\gamma_1 + \frac{5}{4}\,\gamma_2{}^2\,\gamma_2 \quad \Rightarrow \quad \alpha_{r_{\mathrm{rot}}} = \operatorname{arctanh}\frac{5}{37} = 0.1360$$

:seitidipar eseht fo ecnereffiD

$$\alpha = \alpha_{r_{\mathrm{rot}}} + \gamma_2{}^2\,\alpha_r = 0.1360 + \gamma_2{}^2\,1.5223 = 1.3863\,\gamma_2{}^2 \quad \Rightarrow \quad \text{o.k.}$$

.noitator fo elgna emas eht wohs sexa etanidrooc eht dnA

:γ_1 emit fo sixa eht fo noitator citsivitaleR

$$\mathbf{r} = \gamma_1 = 1\,\gamma_1 + 0\,\gamma_2 \quad \Rightarrow \quad c\,t = 1 \qquad x = 0$$

$$\Rightarrow \quad \alpha_r = \operatorname{arctanh}\left(\gamma_2{}^2\,\tfrac{x}{c\,t}\right) = \operatorname{arctanh} 0 = 0$$

$$\mathbf{r}_{\mathrm{rot}} = \gamma_{1\mathrm{rot}} = \frac{1}{8}\,(17\,\gamma_1 + 15\,\gamma_2) = \frac{17}{8}\,\gamma_1 + \frac{15}{8}\,\gamma_2 \quad \Rightarrow \quad c\,t = \frac{17}{8} \qquad x = \frac{15}{8}$$

$$\Rightarrow \quad \alpha_{r_{\mathrm{rot}}} = \operatorname{arctanh}\left(\gamma_2{}^2\,\tfrac{15}{17}\right) = 1.3863\,\gamma_2{}^2$$

:seitidipar owt eseht fo ecnereffiD

$$\alpha = \alpha_{r_{\mathrm{rot}}} + \gamma_2{}^2\,\alpha_r = 1.3863\,\gamma_2{}^2 + \gamma_2{}^2\,0 = 1.3863\,\gamma_2{}^2 \quad \Rightarrow \quad \text{o.k.}$$

The relativisitc rotation of the axis of space γ_2 will be thrilling now.
.won gnillirht eb lliw γ_2 ecaps fo sixa eht fo noitator citsivitaler ehT
.delaever eb lliw nietsniE fo dlrow citsivitaler eht fo ytreporp citsatnaf A

$$\mathbf{r} = \gamma_2 = 0\,\gamma_1 + 1\,\gamma_2 \qquad \Rightarrow \qquad c\,t = 0 \qquad x = 1$$

$$\Rightarrow \qquad \alpha_r = \operatorname{arctanh}\left(\gamma_2{}^2\,\frac{x}{c\,t}\right) = \operatorname{arctanh}\frac{1}{0} = ???$$

.yllacitamehtam denifed ton si ti sa tsixe ton seod tluser A

$$\mathbf{r}_{rot} = \gamma_{2rot} = \frac{1}{8}\,(15\,\gamma_1 + 17\,\gamma_2) = \frac{15}{8}\,\gamma_1 + \frac{17}{8}\,\gamma_2 \;\Rightarrow\; c\,t = \frac{15}{8} \qquad x = \frac{17}{8}$$

$$\Rightarrow \qquad \alpha_{r_{rot}} = \operatorname{arctanh}\left(\gamma_2{}^2\,\frac{17}{15}\right) = ???$$

:niaga ereH
.yllacitamehtam denifed ton si ti sa tsixe ton seod tluser A

.erehwyreve noitator fo elgna emas eht teg ot evah ew tuB
.tnetsisnocni yllacigol eb thgim ytivitaler fo yroeht eht esiwrehtO
.ereh raeppa ton tsum elgna denifed-non A

,sexa ekilemit dna ekilecaps fo snoitator eht fo snoitaluclac eht gnirapmoC
elgna eht fo tluser tcerroc eht evig lliw seulav eht taht deciton eb nac ti
.sixa ekilecaps eht htiw gnilaed nehw degnahcretni era setanidrooc eht fi

.eciohc rehtona evah ton od eW !ti od ew oS

,noitator fo elgna eht dnif ot srotcev ekilecaps htiw gnilaed nehW
egnahcretni setanidrooc ekilecaps dna ekilemit
.noitator fo elgna citsivitaler tcerroc eht mrof ot

sehsifrats taht tcaf egnarts eht fo ecneuqesnoc a si tluser sihT
.emit sa ecaps erusaem dna leef seiticolev lanimulrepus htiw gnivom
.[15] ecaps sa emit erusaem dna leef sehsifrats lanimulrepus eseht dnA

,snoydrat eht era sehsifrats lanimulbus dna snamuh eW
.[15] snoyhcat era sehsifrats lanimulrepus elihw

.emit si ylpmis emit dna ecaps si ylpmis ecaps dlrow cinoydrat ruo nI
sehsifrats lanimulrepus fo dlrow cinoyhcat eht nI
.emit onyhcat emoceb lliw ecaps ruo
.ecaps noyhcat emoceb lliw emit ruo dnA

.srevresbo lanimulrepus rof emit fo sixa na si ecaps fo sixa nA :noisulcnoC
.srevresbo lanimulrepus rof ecaps fo sixa na si emit fo sixa na dnA

.seiticolev lanimulrepus rieht leef ton od srevresbo lanimulrepus suhT
,thgil fo yticolev eht naht rellams seiticolev erusaem syawla yehT
.degnahcretni era sexa rieht esuaceb

.sgnieb lanimulrepus fo seye eht ni snoyhcat eht era snoydrat eW
,lanimulbus dna lamron etiuq leef esruoc fo sngieb lanimulrepus eht dnA
.meht rof stnemevom lanimulbus fo dlrow a si ylpmis dlrow rieht esuaceb

ecaps fo sixa cinoydrat eht fo noitator citsivitaleR   :suhT
emit fo sixa cinoyhcat a fo noitator citsivitaleR =

$$\mathbf{r} = \gamma_2 = 0\,\gamma_1 + 1\,\gamma_2 \qquad \Rightarrow \qquad c\,t = x_{super} = 0 \qquad x = c\,t_{super} = 1$$

$$\Rightarrow \qquad \alpha_r = \operatorname{arctanh}\left(\gamma_2{}^2\,\frac{x_{super}}{c\,t_{super}}\right) = \operatorname{arctanh} 0 = 0$$

$$\mathbf{r}_{rot} = \gamma_{1rot} = \frac{1}{8}\,(17\,\gamma_1 + 15\,\gamma_2) = \frac{17}{8}\,\gamma_1 + \frac{15}{8}\,\gamma_2$$

$$\Rightarrow \qquad c\,t = x_{super} = \frac{17}{8} \qquad x = c\,t_{super} = \frac{15}{8}$$

$$\Rightarrow \qquad \alpha_{r_{rot}} = \operatorname{arctanh}\left(\gamma_2{}^2\,\frac{15}{17}\right) = 1.3863\,\gamma_2{}^2$$

:seitidipar owt eseht fo ecnereffiD

$$\alpha = \alpha_{r_{rot}} + \gamma_2{}^2\,\alpha_r = 1.3863\,\gamma_2{}^2 + \gamma_2{}^2\,0 = 1.3863\,\gamma_2{}^2 \qquad \Rightarrow \qquad \text{o.k.}$$

.[124 .p, 16] emitecaps ni sdrawedis netfo erom evom dluohs ew ,noitseuq oN

:eb tsum noisulcnoc lacihposolihp eht dnA

evitcepsrep emit a fo sisab eht no elbissop ylno era selgna fo snoitatupmoC
.emit fo noitaterpretni lanimulrepus a seriuqer evitcepsrep siht fi neve –

yllacitamehtam detaerc evah ew dlrow eht fo erutcurts eht fo noitadnuof lacigol ehT
.sevitcepsrep lanimulrepus eriuqer ot smees

erutaretiL 16

ot gninraeL ,rehcaeT a ekiL gniknihT :skcaJ .M ymA ,llebA .K ardnaS [1]
:(.dE) llebA .K ardnaS :nI .margorP daorbA ydutS a ni hcaeT
,evitcepsreP lanoitanretnI nA :noitacudE rehcaeT ceneicS
,2000 thcerdroD ,notsoB ,kroY weN ,srehsilbuP cimedacA rewulK
.153 – 141 .pp ,8 .pahc

.thcin se tbig nelhaZ evitageN – Negative Zahlen gibt es nicht :nroH kirE nitraM [2]
978-3-7562-3808-8 :NBSI .2022 tdetsredroN ,dnameD no skooB – DoB
:noitalsnarT hsilgnE
ton od srebmun evitageN – Negative Numbers Do Not Exist :nroH kirE nitraM
978-3-7562-5832-1 :NBSI .2022 tdetsredroN ,dnameD no skooB – DoB .tsixe

a tuoba secnecsinimeR :(.sdE) rengiW .P eneguE ,ulgonusruK .N marheB [3]
,noitide kcabrepap tsriF .cariD eciruaM neirdA luaP :tsicisyhp taerg
.1990 kroY weN ,egdirbmaC ,sserP ytisrevinU egdirbmaC

,cariD luaP fo efiL neddiH ehT .naM tsegnartS ehT :olemraF maharG [4]
.2009 nodnoL ,.dtL rebaF dna rebaF .suineG mutnauQ
:noitalsnarT namreG
sed nebeL enegrobrev saD .hcsneM etsmastles reD :olemraF maharG
.2018 nilreB ,galreV-regnirpS ,egalfuA .2.cariD luaP seinegnetnauQ

,cariD luaP fo efiL neddiH ehT .naM tsegnartS ehT :olemraF maharG [5]
.2009 kroY weN ,puorG skooB suesreP / skooB cisaB .motA eht fo citsyM

etnemoM ehcsitehtsÄ .tseiB sad dnu enöhcS saD :rehcsiF reteP tsnrE [6]
.1997 nehcnüM ,galreV repiP .tfahcsnessiW red ni
:noitalsnarT hsilgnE
ni tnemoM citehtseA ehT .tsaeB eht dna ytuaeB :rehcsiF reteP tsnrE
ecneicS regnirpS. ,noitide revocdrah eht fo tnirper revoctfoS .ecneicS
.1999 nilreB ,aideM ssenisuB +

dna scitehtseA .ytuaeB dna hturT :rahkesardnahC naynamharbuS [7]
,ogacihC ,sserP ogacihC fo ytisrevinU ehT .ecneicS ni snoitavitoM
.1987 nodnoL

.piznirpstätivitaleR saD :(.dE) resegarT gnagfloW [8]
.snietsniE eiroehtstätivitaleR ruz netiebralanigirO nov gnulmmaS eniE
.2018 nilreB, murtkepS regnirpS ,egalfuA .2

.ytivitaleR no srepaP s'ikswokniM .emiT dna ecapS :ikswokniM nnamreH [9]
.2012 cebeuQ ,laertnoM, sserP etutitsnI ikswokniM

eht fo weiV weN A .noitoM fo yrtemoeG lacitpO :bboR ruhtrA derflA [10]
.1911 egdirbmaC ,snoS & reffeH .W .ytivitaleR fo yroehT

.stsicisyhP rof arbeglA cirtemoeG :ybnesaL ynohtnA ,naroD sirhC [11]
.2003 egdirbmaC ,sserP ytisrevinU egdirbmaC

.etueh eiroehtstätivitaleR elleizepS :ikslefaR nnahoJ [12]
.2019 erutaN regnirpS / murtkepS regnirpS

hcubrheL :ztihcsfiL hcstiwoliahciM inegwE ,uadnaL hcstiwodiwaD weL [13]
,egalfuA .9 .eiroehtdleF ehcsissalK .II dnaB ,kisyhP nehcsiteroehT red
.1984 nilreB ,galreV-eimedakA
:noisreV hsilgnE
fo esruoC :ztihsfiL hcivoliahkiM ynegvE ,uadnaL hcivodivaD weL
,noitidE .4 .sdleiF fo yroehT lacissalC ehT .II .loV ,scisyhP laciteroehT
.1987 grebledieH ,notsoB ,madretsmA ,nnamenieH htrowrettuB
,egalfuA .4 .hcubnehcsaT-kitamehtaM :hcsoB lraK [14]
.1993 nieW ,nehcnüM ,galreV gruobnedlO .R
.thgiL naht retsaF snoitoM dna ytivitaleR laicepS :dlognyaF sesoM [15]
.2002 miehnieW ,HCV-yeliW
.kisyhP nenredom red eirogellA eniE .dnalnetnauQ mi ecilA :eromliG treboR [16]
.1995 nedabseiW ,giewhcsnuarB ,tfahcsllesegsgalreV nhoS & geweiV .rdeirF

:koobsiht fo noisrev lanigiro eht si siht dnA

red tätivitaleR eiD
nehcnretseeS

Negative Zahlen gibt es nicht - Die Relativität
der Seesternchen

Martin Erik Horn

.ni detseretni eb thgim uoy skoob erom era erehT
.pohskoob ruoy ni ro <u>ed.bod.www</u> ta meht dnif nac uoY

:nroH kirE nitraM
ILULAP APNAN ANOS
ARBEGLA ILUAP –
2024 tdetsredroN ,DoB

:nroH kirE nitraM
ILULAP APNAN ANOS
DeQ'Im 'Il'wap –
2024 tdetsredroN ,DoB

etoN lanosreP trohS

.em yb sdrawkcab nettirw era skoob emoS
.nognilK ni nettirw era skoob emoS
.em yb anoP neletiS ro anoP ikoT ni nettirw era skoob emos dnA
.yaw taht tsuj s'tI
,nialpxe ton lliw I gnieb emit eht rof ,on dnA
.taht gniod ma I yhw
:siht ylnO
,esrevinu eht fo trap siht nI
yxalag ruo fo larips tnavelerri ,ynit siht ni
sgniht eseht tuoba etirw ot rennam thgir eht yltcaxe si ti
.yaw siht ni